MEETING THE MESSIAH THE THEORY OF EVERYTHING & THE HOLY GRAIL

TR & DK

Published by Resolutions Publishing www.resolutions.world
ISBN: 978-1-7641121-1-6

First edition, 2025

Table of Contents

Chapter 1

The Geometry of Return

The wolf came to me again. On the steps of night's dreams, before I left London for Ireland, soft-footed, watchful, guiding me as it once did when I was a child, with patience, with quiet assurance. I stumbled upward along the trail, my breath shallow, the earth rising beneath tall timber. The trees loomed, not menacing, but like old sentinels: stoic, knowing. We climbed toward a cabin, or something cabin-like, cradled on the slope of a hill. Its large bay window glowed faintly, beckoning, steeped in that peculiar kind of light that seems neither of day nor night, but suspended between like a lantern of ethereal hope, its glow struggling against the weight of the night.

It was a dream I had experienced often as a girl, when the world still shimmered of mystery at the edges. I had not thought of it for years. And yet there it was, the wolf and cabin had returned, more vivid now, more insistent, as if I had been drawing nearer all along without realising. This time, I reached the edge. The cabin, not far. I paused, and so did the wolf. My fingers moved without thinking, threading through its thick, warm coat. There was a silence between us, vast, weighty, and within it, something passed between us. A knowing. I felt its heart, ancient, wild, carrying paths I would never tread. And I knew, with the same quiet surety, that it had long known mine. Then I awoke.

The presence did not leave me, not entirely. It lingered, a kind of quiet company on my return from London to the cottage, as I laced my shoes for a run as the frost bit at the window panes. Before I left,

I wrote it down, the dream, the wolf, the light between us, as if to fix it, however faintly, before it slipped again into the folds of forgetting.

  I wrote what I remembered,
  or imagined,
  or became.

I live in woven packs of warm love that dances upon winter.
We run through raven-black nights, across shivering
valleys of amber.
Alpha breathes in Omega, beneath soft, beautiful,
unyielding strong,
A leader of change, of charge, for the awakening young.
Today, the hunt was long. Now I ramble, full,
Carrying myself, cutting deep tracks, rising toward the hill.
To my moon, unfurl your pale rise beyond dusk, through the
bone of night,
My solace, my mirror, my ancient song of light.
Beneath the whispering timber, the hunters' scent, it comes.
But tonight, my golden eyes are sharper than blade or gun.
Mortals shall lose my padded trail in veils of gathering night,
Through folds of mist and growing mounds of winters white.
No scent lingers. I ascend into the relief, returning to the pack,
Where fear dissolves into the weave of warmth and surrender.
Until we wake again, and roam again, an earnest,
ageless hunt again,
To dance entire across wafers of crystal white, sailing
pure and tender.
The cubs still fumble songs, small howls, sweet, wild,
and uncertain.
Soon enough, they'll scramble the ridges to the crest of
thinning shadows,
Where breath turns ghost-mist, in towering cathedrals of black
timber nights,
Drawn toward you. My pull. My argent tide.
My moon.
My moon.
My crying light.

I set the page down to tinker upon later. Then, without ceremony, I opened the cottage door and the cold morning, angelic in its hush, met me, frost spun delicate lace across the branches, blanketing the grass, all sparkling gently beneath the breath of morn. I began jogging along the familiar path, a ribbon of earth that stretched toward the quiet country road. Mornings like these made starting the motor challenging, yet there was something profoundly invigorating, almost triumphant, in braving the chill to jump-start the awakening day. I cupped my hands, brought them to my lips, and exhaled, the brief warmth a small comfort. What I needed now was to settle into my rhythm, get the blood flowing, and push past the biting cold.

As I passed by the old ash trees, their bare white arms lifted to the softening sky; the sun slipped her golden fingers over the horizon, coaxing the small birds into their morning hymn. I drew deep breaths of the fresh, frosty air, easing into my stride, beginning to warm into it all as old Gilmore's Inn came into view, I then turned left and headed along the coast, welcomed by the rolling hush of the waves as they shuffled pebbles along the shore, like ancient storytellers, murmuring secrets long lost to the wind and sea.

After a few kilometres, I reached Galway Bay and paused, captivated by the endless stillness. I looked out over the ocean, deep green and smooth, while above, the sky stretched out endless and unbroken, a serene clarity that seemed to hold its breath as I gazed at the Aran Islands in the distance. I continued my steady pace along the shore, the scenery unfolding quietly beneath each stride, within each breath, soft spiralling wisps escaped and wandered into the stillness.

The pulse of my heart was proud and steady, around -110 bpm, as I reached my ritual three-kilometre mark, where I was met by a familiar grey mare with a blond flash in her tail and her growing white colt, their quiet presence a steady comfort. They ran alongside me for a moment, as they often did, before I turned back, the rhythm of my steps carrying me homeward. The sharpness of the air, the stillness of the land, it wrapped around me, then evaporated, repelled by a radiant warmth in my quiet sense of renewal.

A while later, I jogged back up the tree-lined driveway, the frost now vanquished by the sun's quiet, mercurial touch. Winding my way uphill through my garden toward my cosy, thatched-roof cottage, I had inherited from my grandparents. Far away but close enough, this place, was a world unto itself, a pocket of untamed beauty where the hills rolled like sleeping giants, and the green fields seemed to breathe with the pulse of Ireland itself. The cottage is my sanctuary; it's where I go to be, relax and escape from the hustle of humanity. It's also where I do my best writing while pondering the shades of life. Around the perimeter of the cottage, there are a variety of old poetic trees, some flowerbeds, and a weathered little greenhouse that provides colour primarily year-round. I love my time here, enveloped in nature and the fondest of memories, away from the chatter of London and the convoluted air and people.

The front of the cottage is south-facing and catches the sun mostly all-day. Around the yard, I have a patchwork of small herb and vegetable gardens I tend to the best I can, but it's always a struggle when I return from afar to wrestle them back into health. The old apple trees in the backyard produce the best apple pies in Ireland. That's what Grandfather always used to say. The cottage has a stone wall encircling it like an ancient vow, where moss and lichen cling on, locked in a silent wedlock. At the front, an old gate holds on with quiet resolve, creaking and scraping as it yields, reluctant, yet faithful. My sister says, "Replace the gate, Sis", but too many fond recollections are there, in that sound locked away, which unravels memories of my grandparents, drawing me back into a golden haze of endless summers. I ran my fingers over the old latch. Strange, the things we hold on to. The weight of memory is sometimes heavier than the thing itself.

I pushed the front door open and stepped inside, to the familiar scent of home wrapping around me. The fragrance of flowers in bloom mingled with it as they perched in my favourite jar, getting their feet wet, glistening luminous greens when the sun's rays angled through the window just right. It was good to be back.

The whitewashed walls and bay windows make it feel airy and bright inside, and there's always a snug fire ready in the fireplace, where I love to sit and watch sparks chase and dance after the crackles. Entering the cottage is the dining room, sunroom and a lounge all in one, and a small study tucked away that could be called my hoarder room. But I see it as my Christmas room, where I discover forgotten treasures tucked away under dust covers, or hidden gems within boxes wrapped in old newspapers. Beyond that, there are two bedrooms and a kitchen. A small set of stairs leads up to a small loft bedroom above, where, as a child, I would gaze out the window, filled with a sense of wonderment...

Entering the kitchen, I rehydrated like a parched sailor sighting land, then got the kettle rumbling for a hearty pot of British char. As I made my way into the living room, there before me lay a dead mouse, with Puddles standing over it, looking at me, then at the mouse, as if he had bequeathed me a gift from the gods. Usually, I set a catch-and-release trap before lightning strikes and liberate them down the way, out onto the grassy green, where there is an old, abandoned barn I call the "Mouse Hotel." This is to avoid Puddles' distinct set of rules, where nothing shall interfere with his cat bowl without dire consequences.

I rescued Puddles from abandonment. It was a morning run in London. It had been raining heavily the night before, and as I returned from a jog, running late for work, I decided to take a shortcut through an old alley to get back to Brit's place for a quick shower. That's when I saw him, a scruffy, mottled brown kitten with a muddied white face and striking emerald eyes. He was sitting in a puddle, looking lost, perhaps a stray, or maybe abandoned by some stray human. I subconsciously needed a companion of sorts to make home homely, and we take comfort in each other in our own unique ways. So Puddles, it seems, was destiny.

Mum takes care of him on my trips away, doting on him, and she enjoys his leisurely demeanour. But there's something about Puddles, a quiet magic, that seems to ease Dad. In the way he instinctively gravitates to him, purring softly, brings a calmness in. Puddles has now become a source of adventure and comfort, not just for myself, but for us all.

Tonight was an early night; I had to squeeze in a small run tomorrow, catch up with Mum and Dad before heading onwards to London, then make my way to Brit's and spend the weekend relaxing before working a few weeks at the museum.

Around 5 a.m., I awoke in a cold, abrupt state before the alarm sounded. I had a dream that was so vivid, clear, indeed a vision it was, so realistic, that it took a few moments for the impact of it all to subside. I lay perfectly still, trying not to disrupt the memory sequence of the dream, recalling it in minute detail as I looked up into the presence of Zeus. I felt dwarfed by the mythical god's enormity as he lowered his arm and pointed his hand to the ground. Upon the length of his arm, I noticed a large, intricate tattoo that looked somewhat like a maze or puzzle, where strange symbols and words were woven between the lines. An elaborate tattoo of a snake ran right down the middle of the maze. Suddenly, I noticed the snake becoming fluid as it started moving downward towards his hand. Through some form of telepathy, Zeus gently tilted his head and communicated to me that the snake represented a key moving towards a specific destination, and once it reached its end, "*all things would become known...*"

My mind couldn't shake the power within the dream as I began pondering over it while getting ready for my short morning jaunt. The writing between the lines of the maze on Zeus's arm reminded me of certain types of ancient coding I'd seen some time ago within the temple of the Aztec imperial capital, Tenochtitlán. The Akashic records also came to my mind. I wondered whether the snake represented something about unlocking doors to access these records.

There are those who believe the Akashic records represent an energy or force that makes up everything in the universe throughout space and time. They call this energy Akasha, a combination of the Sanskrit word ākāśa, meaning hidden or secret space, with the root word kāś, meaning simply "to be."

It also conjured up something from the recesses of my mind, a part of the King James Version, Revelation 20:12-something my grandmother always cracked on about, trying to keep me on the straight and narrow: "*And I saw the dead, small and great, stand before*

God; and the books were opened: and another book was opened, which is the book of life."

So, I jotted down my thoughts and details in my journal and gathered myself for another brisk morning and a short run. After which, I returned for a quick shower, dried my hard-to-manage hair, which, for convenience's sake, requires shortening, and gave Puddles a hearty puff of the hairdryer to persuade him not to use my hair as a climbing pole. I quickly changed into jeans and a T-shirt, shrugged on my parka, and grabbed a quick bite before Puddles and I headed down the way to my parents' house.

My parents live in Galway City, its name drawn from the River Gaillimh, the 'stony river.' Legend tells of Gailleamh, a chieftain's young daughter who was lost to its waters. Stricken by grief, heartbroken, her father is said to have camped by the river's edge, keeping vigil over her spirit, unwilling to leave her alone in its depths. Thus, later, a town sprung up around the location and was called Gaillimh in her honour.

Galway sits on the banks of the River Corrib, which flows into Galway Bay from Lough Corrib. It's there that I have many fond memories of the lake and the fairy-like islands where my grandparents used to take us. My brother Liam, my sister Sinéad, and I would go for long strolls and picnics by the lake regularly when I was young and feisty. I have many fond memories of playing in the water with my childhood immunity to the cold, chasing the ducks for hugs that I could never catch, leaving me, chin up and dog-paddling poorly to the amusement of my family.

As I drove through the streets of Galway City, reminiscing, I couldn't help but admire the artistry in the architecture. The buildings, a blend of old and new, some dating back hundreds of years, whispering to the modern world untold stories of the past. Soon, I arrived at my parents' house and stepped inside, dropping my things off in the living room before setting moody Puddles loose from his carry pack.

My mum was in the kitchen, cooking a sprawling breakfast that would decimate any dietary plan. Her hair was tied in a messy bun, and her blue eyes sparkled as she drew me into her arms for one of Galway's most well-known hugs.

'My sweetie, I'm just getting breakfast on the go. Your father is already at the table,' she said, while Puddles weaved excitedly around her legs, his tail held high like a mischievous wee pirate flag. He knew all too well that her resolve would crumble, and soon he'd be rewarded with a tasty treat, hooking it before it hit the floor.

I stepped into the dining room and sat at the table. My dad was reading the newspaper, but as I joined him, he looked up with a warm smile. His hair remained thick but had begun to grey, and I noticed a few new wrinkles that seemed to have arrived sooner than expected. Yet, he was still the same noble, warm-hearted man I had always known.

'Good morning, my Clover,' he said, reaching out gently to hug me, as if retutning to a rhythm he'd never quite forgotten,

He had coined the name Clover for me years ago, believing me to be his lucky charm, his little four-leaf companion on our treasure hunts through auctions and antique fairs. Perhaps it was my intuitive streak, one inherited from my mother, that led us to hidden gems. Once, I had insisted that he buy an old chest of drawers. He hesitated, scrutinising it for a long moment before relenting. Later, upon closer inspection, we discovered a secret compartment containing a bundle of old Irish Celts' jewellery. Among the treasures was a silver penannular brooch inlaid with lapis lazuli, a stunning deep celestial blue stone. It was one of my favourite finds, a testament to my knack for unearthing hidden beauty, and a story my father loved to retell to his fellow adventurers.

There was something about the antique shop I couldn't quite shake off, missing it deeply, the gentle chime of the doorbell whenever someone wandered in, some with a specific treasure in mind, others just enjoying the space. It was there, in the midst of those forgotten relics, that I learned more than I ever expected. Each object carried its own magical story, and through my father's hands and broad knowledge, I began to understand the life behind every piece. That quiet, undying passion for the past, has stayed with me ever since...

Mum soon arrived with breakfast, balancing three weighty plates, while Puddles trotted closely behind her, in eager anticipation. I reached for the teapot, pouring us all a cup of tea. My father lifted his cup, his hands trembling slightly as he tried to steady it. Parkinson's had been gradually chipping away at him over the years, the reason he had to let go of the antique business. With me as the curator, Liam settled in Dublin, and Sinead was deeply rooted in the old pub. Dad's once-strong body had silently betrayed him. The hardest blow of his diagnosis had not been the physical decline, but the loss of his beloved shop, a cornerstone of his identity and a cherished part of all our lives.

Mum refilled our teacups, a delicate blue china cup rattled on Dad's saucer before he managed to lift it to his lips. She sat beside him, resting a gentle hand on his shoulder, offering him the same quiet strength that had steadied our family for years.

'How are you feeling then, Father?' I asked softly.

'I'm alright,' he replied, a wry smile playing on his lips. 'Just proper weary, girl. Otherwise, this ancient K9 would race you up the way to the local for half a pint. But now, if you gave me a day's head start and cabbie fare, I think your old man might manage a nobel second.'

I could sense, beyond his ever-present humour, how much he missed the shop. He had loved his daily strolls to his second home, chatting with customers, listening to their stories, and sharing his knowledge. The shop had been filled with treasures, intricately crafted coins, old jewellery, and books dating back centuries. It was his sanctuary of history where he could lose himself in the echoes of the past. Mum's beloved climbing roses framed its old charm entrance, their fragrant blooms brushing against large windows that showcased an ever-changing collection of curiosities. Inside, dark wooden floors and furniture stood as a wonderful testament to the passage of time. To my young eyes, the books stacked on those shelves had looked like magic tomes, promising endless adventure.

My father had an incredible eye for hidden gems, often with me, his wee clover, by his side. He'd let me sift through boxes of what others might call "junk," helping him restore forgotten pieces to their former glory. Whenever I stumbled upon something special, I could still picture his signature sideways look, a mix of amusement and inspiration. His love for antiques had clearly shaped my own path, leading me to where I was now, working at the museum, carrying on his legacy in my own way.

'Maybe we can all take a road trip to Dublin and visit Liam when you return from London, eh?' Mum suggested, scratching the bridge of Puddles' nose, his undeniable sweet spot.

After breakfast, I helped Mum clear the plates before settling down with Dad for one last cup of tea. Puddles nestled in his lap, as if sensing the calm his presence brought. I remembered when Mum and I had first noticed the changes, the slight tremor in his hand, his voice becoming softer and hoarser, the occasional flashes of frustration that we had initially dismissed as a sign of age. Then came the unsteadiness in his steps. The doctor had confirmed Parkinson's, preparing us for the road ahead, the speech difficulties, balance issues, and even struggles with something as simple as swallowing. Yet through it all, my father remained steadfast, never losing his humour or resilience.

He once told me, "Girl, live life so that you'll be missed when you're gone." A mantra I have carried with me ever since.

The afternoon drifted by into a blend of laughter and tales, each moment filled with the warmth of shared memories, catching up on the bits of life we'd missed. My sister, always knee-deep in her latest project, this time, the renovation of the old pub she'd bought years ago, had us all in fits with her latest escapades. But as it often does, time seemed to slip away, and before we knew it, the hour had arrived for me to leave and catch my ferry.

Hugs all around, and Mum walked me to the door, seeing me off with her usual warmth. 'It's so good to have you around,' she said. 'See you soon enough then, and Puddles will be twice his size on your return.' She smiled, pressing a kiss to my cheek.

I was already looking forward to spending more time with them. Road trips, taking shape in my mind. I loved excursions where the roads weaved with the trees. And who knew? Perhaps we'd even stumble upon a few hidden treasures along the way.

Chapter 2
A Thread Toward Familiar

Climbing into the car, I eased my way toward Galway's city centre, often called the 'Venice of the West' by Yeats himself.. The streets, with their cobblestone charm, were like lyrical verses lined with old-world shops and cosy cafés. It was the kind of place that called to a wanderer's soul, offering the quiet promise of unexpected adventure.

I parked, climbed out, stretched, and let the city tug at me. I wandered, half lost, from shop to shop, eyes grazing over the oddities in the windows, pausing now and then to listen to the buskers. Whose melodies threaded through the streets like a mingling wind of song. One of them caught my attention. Mick. A familiar busker from Dublin. Mick had that rare kind of voice, the kind that could elevate rocks with his vocals, his slightly gravelled voice enchanting as ever. Rough-edged but rich, like an old cobblestone road that you'd walked before with a wistful friend. His original songs carried weight; one about the homeless hit deep, deserved more ears than it ever got. But as I got closer, his tune shifted. He began singing *The Town I Loved So Well* by Phil Coulter. And just like that, the whole street seemed to pause, caught in the quiet spell of a song that knew its way through history, longing, and loss.

I sank onto a nearby bench and listened to Mick singing, his voice sweeping me away on a beautiful tide of emotions. I closed my eyes, letting the music carry me back through the years of growing up with these old songs. Towards the end, I opened my watery eyes, and he caught my gaze, smiling softly. I walked over and dropped a few pound notes into his case.

As he sang the final verses, words of loss, and what is lost forever, resilience, and the hope for a brighter tomorrow. The weight of history echoed in his voice, carrying the past forward, unbroken through time.

'Thank you kindly,' he said, before continuing with his rendition that moved me deeply...

I wandered a little longer, letting the city's rhythm weave its magic around me, the smell of fresh bread rolling out a bakery's door, mingling with the vibrant chatter and laughter spilling from the cafés and pubs. It was that unmistakable feeling of coming home. Galway, with its storied past and literary soul, has been home to some of Ireland's most famous writers, W. B. Yeats and Lady Gregory among them. And yet, despite its rise as a popular tourist haven, it has held onto its quiet, small-town essence, as if time itself slows to savour its beauty.

A week earlier, I had booked a train ticket from Galway City to Dublin, an act of quiet precision, a small ritual of order in an otherwise unpredictable world. Being the sort who prefers her time as an ally rather than an adversary, I had given myself a generous window to reach the station. The air, crisp with the lingering chill of spring, carried the city's murmurs, shuffling footsteps, the occasional hum of conversation, the low groan of engines idling in the distance.

The station received me with quiet indifference, as I settled in to wait before departure. The station itself was unassuming, a structure built for function, its muted tones blending into the geometry of the city as if it had always been there, neither intruding nor demanding notice. But for me, it held ghosts, memories rising unbidden, curling around the edges of my mind. Childhood journeys. The scent of train oil and worn leather seats. The gentle anticipation of going somewhere. And now, here I was again, caught between places, waiting.

I was eager to return to London, where the museum was abuzz with preparations for a new project I couldn't wait to dive into. Yet, the thought of leaving Puddles behind, tugged at my heart. There were times I managed to lug him along from home to London, but this trip was too tangled with side errands to make it possible. Besides, I knew Father would find comfort in Puddle's quiet, yet mischievous presence, and Mum, ever the doting caretaker, would fuss over him no end.

My stretched ride sighed into Galway station, its arrival marked by a rhythmic clatter that felt both reassuring and perfectly on time. I stepped aboard, weaving through the near-empty carriage, where a few solitary travellers sat cocooned in their own quiet reveries, lingering in their own silent worlds. My window seat awaited; sliding into my spot, I settled my bag down beside me, and fished out my journal, eager to capture wandering thoughts worth penning before they took flight beyond memory. With a gentle lurch, the train slipped free of the platform, and I settled in, awaiting landscapes to begin breezing by in a symphony of familiar shapes to unfurl in a blur of familiar ghosts and shifting light... There is a quiet magic in train journeys that I enjoy on these round trips between home and London.

The train pulled and shuddered forward, releasing me from the city in a slow, measured retreat. The city fell behind, its hum giving way to the hush of fields and crooked lanes, where old cottages leaned into the dusk like tired storytellers, still warm with the breath of a thousand forgotten afternoons. Beyond them in the green, the land stretched wide, fields rolling endlessly, speckled with grazing sheep and the occasional ruin of some long-forgotten stronghold. A familiar stretch of land rolling by was a soothing sight, something deeply familiar. But as my eyes followed the sweep of the land, a quiet sadness crept over me. The trees, or rather, their absence, left the view feeling oddly bare, like a story missing its middle, like a song with a missing note. The emerald fields still shimmered in their defiant brilliance, but the deep, endless canopies that once made Ireland the Isle of Wood had faded into something more spectral, more imagined than real. What remained was a whisper of what once was, a land half-remembered, its dense woodlands reduced to memory.

As we pulled into Dublin's Heuston Station, I gathered my things, ready to hustle my way into the city's heartbeat. Heuston was grand, it stands like a relic of another era, its Italianate façade a whisper of grandeur in a world that no longer pauses to admire such things. Corinthian columns rise in quiet defiance, their symmetry unshaken by time, while carved swags and urns sit atop the attic storey like forgotten heirlooms.

As I stepped out of the station, the shift was instant, a jarring contrast to the rolling quiet of the countryside. Cars and buses zipped by in chaotic rhythms, footsteps beat their own hurried percussion against the pavement, blending in with the cacophony of urban life. The air, thick with exhaust, carried the unmistakable scent of Dublin, a sharp contrast to the fresh, earthy calm of the countryside I had just left on the train.

Liam, predictably late, arrived with all the flair of a brazen urban ninja. He swung into the street in his Taekwondo gear, window down, grinning like a rogue. "Move it, Sis, before the boys in blue throw me in the back of the paddy wagon!" I sighed, hoisting my bag higher as I wove through traffic like an overburdened mule, dodging taxis and cyclists with the grace of a collapsing deckchair.

'You're still in your pyjamas? Did you fall out of bed to get here?' I teased, eyeing his martial arts uniform.

Liam laughed. 'Nah, Sis, the other instructor came down with Covid, so I'm filling in. Sorry, D, only got time to drop you off before I hustle back to work.' Always the whirlwind, my brother.

Liam was consistently inconsistent, yet somehow, he managed to juggle everything. How he does it still amazes me.

'So, how's the old man?' he asked.

'He's doing well, considering. Hasn't lost his spark, and Mum's propping him up like a soldier in arms.'

Liam grinned. "Ah, that's good to hear. I'll get up to see them in a week or so and take them up to Sinead's place for a night of singing.'

'I'll get back to them in a month after a stint at the museum,' I said. 'A lot going on in London.'

Liam chuckled. 'That cat of yours will be a fluffy fat ball when you get back. You know how Mum gives in. You'll have to take him to bacon rehab when you return.'

'Yes, I'm prepped for that,' I said, smiling. 'I've got his favourite toy hanging on the treadmill and will crank it down to slow burn, letting him walk it off.'

A bit of craic, and before too long, we were at the drop-off point. A quick hug, another swift exit, and a lumbering donkey run to the safety of the sidewalk. Liam had given me a solid workout.

I bundled myself into a cab near Dublin Port, located on the River Liffey, about two kilometres from the city centre. The port was a hive of activity, with ships coming and going while towering cranes loaded and unloaded containers. As I saddled along, the surrounding area was abuzz with life, restaurants, bars, and shops spilling out onto the streets, their neon signs flickering in the evening light.

The taxi rolled to a stop at the ferry boarding area, and I paid the driver before making my way to the ticket office. A brisk exchange later, I had my passage secured to Holyhead, Wales. With bags in tow, I wove my way to the waiting area. Soon, boarding was announced, and we moved as one towards the gangway. My group was among the first to board. A low murmur of voices drifted through the air, families herding children, lone wanderers lost in thought nursing their paper cups of coffee. The ship loomed above us, a steel colossus, its decks sprawling wide, its weathered hull etched with the wear of countless crossings.

All aboard, as I found my window seat and slumped, watching as we pulled away from the dock and out toward open water. As we eased away from beyond the bay, I then decided to stretch my sea legs. Strolling the deck, I marvelled at the vast expanse before me, that stretched out in every direction, a restless, glistening thing. The Irish wind absorbed me like an old song half-remembered, and in that moment, like many others, it felt as if my mind wielded a brush and canvas. Painting ephemeral scenes that would live on within me, fading but never quite gone.

The Irish coastline never failed to stir something in me, a quiet reverence, a wild pull, the cliffs standing defiant against the thrashing wind, the hills rolling back in ancient, battered folds. The breeze sharpened, snapping at my coat, and the waves grew fierce yet poetic. I found myself jotting a quick line in my journal: *on their maned crests ride the powdered bones of sailors, souls once claimed by the talons of the sea*. Tucking the notebook away, I ducked inside, drawn by the promise of warmth, where the duty-free shop waited: shelves lined with the familiar, a good bottle of red for Brit, and a top-shelf box of chocolates, our unspoken tradition.

Hunger crept up on me as I wandered, and I soon found myself in the cafeteria. A simple sandwich and a cup of B-grade coffee sufficed, though neither was particularly inspiring. My mind drifted to Brit. A small smile tugged at my lips as I thought of our many adventures- and the countless yet to come. She was one of the best friends anyone could hope for, a rare and unwavering presence in my life. To know her was a gift, the kind of friendship you clasp with both hands and never let go.

We had met once upon a time at Oxford, bonding over a shared love of books, history, and travel. Brit had been studying medicine while I had immersed myself in archaeology. Our friendship was stitched together by our mutual fascination with things beyond touch, the weight of time, the echoes of history, and the stories woven into the pages of great literature.

Graduation had come too soon, time splintering as we stepped across the stage into the unknown. Determined to seize the world before it slipped from our grasp, we had thrown on our backpacks and set out across Europe, walking its streets as if we could capture it all, every moment, every city, every story. That had been eight years ago. The world had changed in ways we had not foreseen, yet we remained, bound by the kind of history that makes people inseparable.

Brit, ever the rebel, had forged her own path. She had turned away from medicine, realising it was not her calling, and found herself drawn to horticulture. Her parents had grieved the change as though she had forsaken the life they had envisioned for her. But she had built something remarkable, something that belonged entirely to her. Over time, even they, with their quiet disappointment, had come to respect her decision, perhaps even envy it, in the way people envy those brave enough to heed the voice within.

Her business, a delicate balance of landscaping and interior design, was her soul's work. She was never one for complacency, always pushing, always evolving. That defiant spirit had always been a force to reckon with, and I, grateful beyond words, had been beside her through it all-through every journey, every moment, as we charted the ever-unfolding course of our lives.

The ferry eased into Holyhead, the slow, deliberate pace of docking mirroring the gentle rocking of the waves. I stood at the rail, breathing in the crisp, salt-tinged air. Once ashore, I made my way beyond the terminal gates, my steps unsteady from the hours at sea. Holyhead, a quiet Welsh town on Holy Island, bore the weight of history, its past etched into the ruins of a 6th-century monastery and the remnants of a 12th-century castle. I wandered through its streets, stopping at a modest café for a quick bite and a proper barista's choice coffee, before making my way to the train station.

The journey to London stretched over five hours, the countryside shifting as we pressed eastward. The rolling green hills gave way to the suburban outskirts, and soon, the skyline emerged, the unmistakable silhouettes of Big Ben, the Houses of Parliament standing firm against the misty sky. London, timeless and unyielding, unfolded before me like the opening lines of a well-loved novel.

Stepping onto the platform at Euston Station, I navigated the bustling crowds, my thoughts already turning towards Brittany's flat in Chelsea, a gift from her well-heeled parents. Independent to her core, Brit had initially refused their offer, but a compromise had been struck where she would pay them back, over time. A hefty sum, no doubt, but one she shouldered with her characteristic determination. Her parents had relented, trusting her to find her own way, balancing her repayments without denying herself the little indulgences she so delighted in.

Her flat was more than just a home; it was a sanctuary, a reflection of her soul. She had insisted that the spare room be mine whenever London called me back, and I had embraced it, contributing where I could. Her generosity was boundless, a quiet force that always left me humbled.

When she had first moved in, she had thrown herself into renovations, dragging me over from Ireland to assist. "We start with your room," she had declared, armed with paint samples and a determined glint in her eye. After much deliberation, we had settled on a plan, rolling up our sleeves and setting to work with a Faux Venetian plaster finish. Half a bottle of red had fuelled our creativity, and when the final strokes had been laid, we collapsed onto the floor, two dishevelled artisan artists surveying their masterpiece.

The walls, a weathered white with sweeping strokes of sandy sky blue and subtle hints of tan, evoked the dreamlike quality of a Monet painting. The light from the bay window cast shifting hues across the textured surface, lending it an almost ethereal charm. It had reminded us both of Monet's Woman with a Parasol, a quiet homage to the Impressionists we both admired. As we sat there, the weight of paint on our clothes, the warmth of wine in our veins, we were wholly content. The room had surpassed our imaginings, a space alive with movement and light. We had been utterly chuffed.

So then, we found ourselves lost in laughter, dusting off memories of the time I first took on the cottage and the notion to renovate it. Brit, ever more adept in the hands-on department, was easy to coax to the scene. She arrived brimming with enthusiasm and ideas, a force of nature ready to transform the place. When Brit gets motivated, there's no stopping her until she crosses the finish line, often with some unexpected gusto.

On the second day after she arrived, I ventured out for supplies, leaving her to her own devices. Upon returning, I was met by an indoor geyser erupting within, thanks to pipes beneath the sink being wrenched loose in her zealous dismantling of the kitchen.

What followed was pure chaos:

The two of us scrambled about with buckets, drenched in our bib and braces, looking like ten-year-old girls in a right panic, slipping and sliding about in our wellies like ducks on a frozen pond. It became a frantic relay from the kitchen to the outdoors, water sloshing, buckets brimming. Between laps, I'd stop to wrestle with the stubborn kitchen window for a faster way to expel the water, but it resolutely refused to budge. Even Brit gave it a go, attacking it like a mad crook with a small pinch bar. That window had been obstinate for years, choosing this moment of crisis to remain firmly shut.

Brit eventually stopped, huffing and puffing, and barked at me to go outside with a spade and wrench to find the mains tap. I took her command and set about frolicking out front with the grace of a drunken badger, clawing at the earth like a prospector convinced she could tunnel her way to a better life, or at least to the water mains. All the while, Puddles sat perched on the dry windowsill, watching the spectacle with a bemused feline fascination. His head swivelled back and forth as though taking in a tennis match, observing Brit sliding around the kitchen and me frolicking about in the great outdoors, spade and wrench in hand, desperately trying to find the mains tap. It was around then. I started breaking into fits of laughter, realising things couldn't possibly get worse.

As Brit Brit hollered from the door., 'D, for the love of all things holy, find the fecking thing! Or I'll be doing a faffing water ballet in the next five minutes!'

We emptied ourselves of unbridled laughter that day, as if the whole world had briefly forgotten to be serious. Eventually, I found the tap, silencing the geyser and restoring a semblance of calm. By then, we were both soaked to the bone; I was muddied from head to toe, shivering and half-spent, and yet, the memory of that splendid little calamity has never failed to summon our laughter.

In just over three weeks, the cottage was transformed, its character preserved yet much brighter and fresher. But I don't think I've ever felt so worn out trying to keep up with someone. Brit, of course, was tireless and had to be talked down from fully landscaping both the front and back gardens. I, however, had reached my limit.

'Let's park the bus here,' I told her, slumping against the doorframe. 'Draw up a plan, it'll give me something to chip away at.'

She shot me that mischievous smile and teased, 'You slacker! But I suppose you could beat me in a 6k run, so I'll let you off.'

After catching up, we made the easy decision to head out for dinner and drinks at our favourite spot in Soho, the kind of place where the atmosphere wraps you in warmth, and the cuisine begs to be indulged. We went full steam ahead on the menu, at my expense, naturally. A ritual we've refined over the years during my regular returns to London. After a delectable evening out, we waddled our way back home, full and content. Once inside, we slumped onto the couch, put on some music, and let the hours slip by in conversation. Eventually, the day caught up with my body, exhaustion rolling over me like a slow, slumbering wave.

With one eye closed and the other half-open, I shuffled off to my room and collapsed onto the bed, half-undressed.

Lights, out.

Chapter 3

Dream in the Shadow of Nations

The next morning, I rose early, pulling myself into the day with a slow, reluctant stretch. A light jog through the crisp air worked its way through the last dregs of sleep, the cold biting just enough to wake me fully. A quick shower later, and I was freshened up and ready for the museum. The hours unravelled in an unhurried rhythm, conversations ebbing and flowing, catching up with colleagues, sharing whispers of reverence for the artefacts before us. Each piece, a remnant of hands long vanished, holding the weight of lost worlds lingering in the quiet, waiting to be uncovered.

By late afternoon, I found myself at a nearby café, a refuge I cherish for its excellent coffee and peaceful hum. Settling into a cosy spot by the window, I spent some quiet time writing in my journal, nibbling on a generously-sized cookie, and sipping a tall latte. The vision of Zeus, still oddly swimming around in my subconscious, surfaced again. I scribbled a few more thoughts down, trying to untangle its meaning, before tucking my journal away and setting off back to Brit's flat.

On my way there, I bumped into Paul, Brit's partner. We approached her door from opposite directions, and the moment he spotted me, he flung his arms wide like a dancer mid-routine in Zorba the Greek.

'Bring it here, madame D, bring it!' he called in his thick English accent brimming with a theatrical French tone.

I couldn't help but laugh as we shared a warm embrace, his customary double kisses landing on both cheeks. Paul is, in every sense, a kaleidoscope of charm and vibrant energy. As a struggling artist, he spends his days working in an art-store gallery, but his evenings and weekends belong to his own creations. His work leans towards abstract and expressionism, though his talent stretches far beyond encompassing, with a sharp skill for sketching and an eye for still life. Slowly but steadily, his name is finding its way into circles that appreciate his unique artistry.

'Let's go in, shall we?' he said, his voice enthusiastic and buoyant as he bounced up the steps.

Brit was already inside when we arrived, so we all settled down to share tales of our day's adventures. Brit recounted her struggle manoeuvring a wheelbarrow with a flat tyre through a labyrinth of stonework, a feat she described with great theatrics, complete with hand gestures. I filled them in on the museum's exhibitions in the making, and soon, the conversation turned to Paul's work and his ongoing art.

Paul, ever the cheeky one, turned to me with a slight grin. 'So, D, when shall we do your naked portrait then?' he asked, raising an eyebrow.

Before I could respond, Brit jumped in. 'Not until you can do my naked portrait without making me look like a lorry driver!' Her voice, playful yet pointed, filled the room.

Paul, adopting a mock French accent, retorted, 'Oh, bumpkin, you know I aim for the essence!'

'Essence?' Brit shot back, her tone dripping with mockery. 'You gave me the essence of a Geordie who could lug two full kegs of ale up a flight of stairs!'

That did it. I snorted mid-sip, coffee spilling as laughter escaped before I could recapture it.

As the evening unfolded, Brit and Paul eventually headed out to catch up with friends, leaving me to enjoy a quiet night in. I had an early morning ahead, so I decided to stay back. I cooked up one of my favourite dishes, pasta with butternut squash and parmesan sauce- a comforting meal that never failed to hit the spot. Afterwards,

I spent some time researching for work, then indulged in a YouTube binge on archaeology. By the time I finally turned in, I was snug beneath my Venetian wall art, the distant hum of the street lulling me to sleep.

That peace didn't last long, though. Around 1 a.m., I awoke abruptly, the remnants of a vivid, life-like dream swirling in my mind. Fumbling, blurry-eyed under the bed, I found my journal and quickly began jotting down every detail I could recall. The dream, clear and strangely urgent, demanded to be penned before it slipped away.

In my dream, she came to me, a strikingly beautiful Native American woman, her presence luminous and otherworldly. She spoke with calm urgency, her voice carrying the weight of something far beyond me. 'You must learn this dance,' she said. 'It is of great importance.' She explained that history was vanishing, images, writings, all traces of it quietly being erased.

She held up a photograph, its edges soft and worn with age. The picture must have been at least fifty years old, yet she looked untouched by time, her face as youthful as the one before me. 'Many are being contacted,' she continued, 'each chosen to preserve something sacred.' Then, without another word, she began to move.

Her motions were deliberate, flowing with a quiet grace. Her gaze, sharp and unyielding, caught each misstep before I even realised it. Again. And again. My body resisted, stiff and uncooperative, until, somewhere between repetition and surrender, something gave. And then, at last, it clicked.. My body understanding in ways words never could.

When I woke, her image clung to me, but more vivid still was her dress. I knew it instantly, it was a Plains-style ghost dance dress, circa 1890, like the ones I'd studied during my time at the Metropolitan Museum of Art in New York. Yet this one was singular. Amid the familiar motifs, a half eagle stood out, its wings poised mid-flight, as though caught between two worlds.

Years ago, I'd fallen in love with them, the dresses, the shirts, drawn to their beauty, their quiet power and hues. There was grace in them that was quiet but undeniable. Over time, I had sought them out in collections, running my fingers over soft, timeworn buckskins, their natural dyes still whispering stories in ochre and indigo. Each design carried meaning, a language written in pigment and stitch, one not spoken but felt.

Handprints, bold and arresting, told of struggle and purpose, red for battle, green for harmony with nature. Stars and moons seemed to glow alongside the crows, eagles, and magpies, messengers to the spirit world. Turtles, the earth's steadfastness, while circles whispered of the unbroken cycles of life. Thunderbirds, fierce and triumphant, a promise of renewal, rising beyond defeat like phoenixes.

Each garment was a masterpiece, achingly beautiful and deeply alive. Yet they bore the weight of tragedy too, their existence a quiet rebellion against the forces that sought to erase them. These works of art were more than relics; they were the keepers of stories, guardians of a wisdom that refused to disappear, no matter how much the world tried to forget.

From the arrival of Europeans in 1492 to the mid—1900s, Native American populations endured devastating losses. A series of tragedies. War, disease, and forced displacement tore through their lands, leaving scars that ran deeper than history liked to remember. Some call it genocide now, too late to change what was done. The slaughter of the bison struck another cruel blow, ripping away both sustenance and spirit. And then came the dismantling of families, institution by institution, as if survival itself had to be stripped from their hands. The injustices are endless, each loss stacking higher, like stones on unmarked graves.

Forced onto reservations, stripped of roaming the lands, and severed from a vast history built around harmony with nature, the Native peoples turned to the Ghost Dance. This sacred dance became their final hope, their last desperate gasp to restore the old sacred ways. It originated among the Northern Paiute people and was brought forth by a spiritual leader named Wovoka, who spoke of a prophecy, a vision of the end of colonial expansion and the return of peace and balance.

The Lakota, their dreams shaped by millennial hope, adopted a variation of Wovoka's teachings. Their movements became an invocation, a prayer etched in dust and memory. It was beautiful and haunting, and for those who watched with fear instead of understanding, it was seen as a threat. And so, it led to Wounded Knee...

I remember reading. Bury My Heart at Wounded Knee when I was young; its pages were heavy with sorrow. At Wounded Knee, the dance came to a final, terrible halt: 153 Lakota were slain, most of them women and children, alongside 25 U.S. soldiers, many of whom died due to friendly fire. It was not simply a massacre, but a silencing, a smothering of a people's last breath to exist as they had once been.

The massacre was more than a tragedy; it was almost the death knell for the Native American way of life, a last stand drowned in bloodshed and heartbreak, marking the end of a way of being, roaming the plains like an ancient wind.

As an Irish person, I have often felt the sorrow tied to this history. So many of my people fled famine, oppression, and brutality to seek refuge in America, carrying with them the scars of those who had held power over us. Yet, in escaping their suffering, they became witnesses, and participants, in the suffering inflicted on the Native peoples who had lived in relative harmony upon the soil, long before my ancestors set foot upon it.

Floyd Westerman, known as "*Red Crow*," a descendant of the Dakota people, may he rest in peace, once spoke of tiocfaidh ár lá, a Gaelic phrase meaning "*Our day will come.*" He interpreted this phrase in a spiritual sense, drawing deep connections between the struggles of the Irish and Native American peoples. Floyd shared a poignant moment when Gerry Adams, an Irish political leader, expressed a desire to apologise to Native Americans, acknowledging a painful history: Sadly, the Irish had played a significant role in the deaths of Native Americans during the expansion of the American frontier. Many Irish immigrants, serving as frontline troops in the King's army, were complicit in the violent spread of colonialism.

Yet, Floyd observed how these tragic histories intertwined, and spoke of the shared resilience of both peoples, and the beauty inherent in the Irish spirit. His reflections were a powerful reminder of the extraordinary human capacity for forgiveness, hope, and healing in the face of enduring hardship.

History has a way of circling back, its echoes carried on unseen threads. In the wake of the COVID-19 pandemic, the Irish people raised nearly $2 million to aid the Navajo and Hopi Nations, where food was scarce, and clean water scarcer still. It was more than charity; it was a gesture steeped in memory. A reflection, perhaps, of 1847, when the Great Famine gripped Ireland, claiming over a million lives and forcing two million more across the sea. And in the midst of that sorrow, the Choctaw Nation, no strangers to suffering themselves, sent $170, worth around $5,000 today, to help feed the starving. A kindness that spanned oceans, refusing to be forgotten.

I often find myself reflecting on the late John Trudell, another remarkable Dakota descendant. His eloquence and intellectual depth made him a voice of clarity and strength in the face of injustice. A political activist, poet, musician, and speaker, Trudell's poem Crazy Horse, set to music, always moves me deeply. In one of his speeches, Tribes of Europe, he shed light on the shared history of oppression endured by the tribes of Europe. He spoke of the Inquisition, a terror that began in France and spread throughout Europe and beyond, a brutal reign lasting over 500 years, with historians estimating that millions perished in tortures and executions..

Few institutions have preached love, kindness, and compassion while wielding power so mercilessly. The Catholic Church, in its long reign, often reserved 'love thy neighbour' for those who bent to its will. Dissent was met not with mercy, but with iron, torture, execution, and the seizure of lands to swell the coffers of the powerful. Cloaked in divine authority. It turned suffering into currency, justifying the hardships of the poor while amassing wealth and dominion.. In doing so, it cast aside the very teachings of Christ it claimed to uphold, abandoning those it was meant to serve. For centuries, this betrayal ran deep, its wounds not only etched into history but into the spirit of those left in its wake. The barbarity of this heresy against love and kindness raged for centuries, leaving a scar on humanity's collective conscience.

Such brutality, borne from European perceptions far removed from the reality of human existence, helps explain why so many immigrants, fleeing their own tyrannies, carried with them distorted views of humanity. The tribes of Europe, once diverse in culture and belief, were indoctrinated into a single dominant tribe shaped by systems of power and hierarchy. This consolidation of identity and ideology fuelled atrocities committed against Native American tribes and other Indigenous peoples across the globe. While we cannot undo the wrongs of the past, we must strive to understand the forces that perpetuate such cycles of violence so that we may break them. Even today, those cycles persist, cloaked in militarism and a warped vision of progress, a narrative designed to serve the interests of those who hold power and wealth and those who desire it.

Human history offers us lessons, but far too few seem to learn them.

Where do we begin? Perhaps with the Mongol Empire, under Genghis Khan (1206-1368), whose sweeping conquests stretched across continents, creating the largest contiguous empire in history. His empire was built on the foundation of relentless violence, a legacy of destruction and bloodshed that left indelible marks upon the earth. Or, perhaps we shift our gaze to the British Empire (1583-1997), with its influence far-reaching, leaving colonies in its wake. Yet what lingered longer was a history marked by violence, exploitation, and cycles of subjugation. These patterns, entrenched over time, echo through the ages, shaping the world in ways still felt today. Beyond the rivers of blood, the suffering that spread so far and wide lies a deeper history, a history of slavery, one that is inseparable from all we have come to know.

The word "*slave*" carries the weight of history, its roots tracing back to the Slavic people, or Slověne. Transformed by the tongues of Byzantine Greeks into sklábos and ésklabēnos, it evokes a distant yet lingering chapter when Slavs were torn from their homelands, their lives bartered as commodities under the relentless will of others. By the 9th century AD, many had been trafficked into servitude by Muslim rulers in Spain, and so began an enduring legacy of human exploitation that cast its shadow across centuries.

Slavery, the grim act of owning and dehumanising another, forcing labour without recompense, threads itself through the annals of human history, murmured in the verses of the Bible and the Quran. At its heart lies a stark truth: the abuse of power. It is a system in which dominion is wielded, reducing lives to nothing more than a means to an end. This monstrous injustice seeped across continents, swelling into the transatlantic slave trade, where Africa became both the source and the sorrow of a trade that devoured millions.

In this dark chapter, power dynamics within Africa itself played a pivotal role. Dominant African leaders and kingdoms, often seeking to consolidate their own wealth and influence, captured and sold their rivals and others into slavery, fuelling the demands of European traders. The continent became entangled in a web of exploitation, with its people corralled and commodified, to satisfy an insatiable market. The devastating impact of this trade left scars on the cultural, social, and demographic fabric of Africa, scars that persist even today.

In the Americas, the scale of slavery unfolded with harrowing contrasts. In the Caribbean, plantations often enslaved 150 or more individuals, their existence tethered to unrelenting toil. In places like Dutch Guiana, Brazil, and the West Indies, mortality rates soared so high and birth rates plummeted so low that populations could not sustain themselves without a steady influx of enslaved individuals torn from Africa's shores.

The United States, though no less culpable, bore its differences. Here, while death claimed many, the birth rate rose, a bitter irony that lessened reliance on the ships that continued to carry human cargo. Still, whether sustained by importation or reproduction, the machinery of exploitation churned on, each system steeped in suffering, each marked by the indelible stain of human cruelty.

I lay there, my mind adrift on an ocean, pondering everything that had surfaced from the depths of my unconscious as if seeking checks and balances. These thoughts swirled around my head and heart, flowing until it was time to get up. This dream had affected me on a deep level I hadn't experienced before. So, as I waited for the coffee machine to reach temperature, I began performing the dance I had been instructed to learn, though I was unsure why...

'What on earth are you doing?' Brit's voice broke through the stillness, her figure framed in the doorway, looking as though the night had left its mark upon her.

'I think it's the Native American Ghost Dance,' I said, attempting a smile for my dishevelled companion. She leaned against the doorframe, eyes narrowing in amused disbelief.

'Well, Pocahontas,' she said, 'make me a double shot. I've got a nasty night after, flu, and I need something to make it go away.'

Mugs in hand, we slipped out onto the balcony, the world stretched out quietly beyond the gates of spring. The flowers were bold in bloom, a quiet herald for the slow shift toward summer. We settled into the old cane chairs, each of us consumed with our own thoughts, yet together, silent and companionable.

'So,' Brit began her voice drowsy, with the remnants of sleep and illness, 'what's the Ghost Dance all about, then? D, you know, my mouth's saying that, but my head's not really in any shape to listen.' I laughed, as always, appreciating Brit's straightforward approach, her honesty unclouded by pretension.

Together we sat in comfortable silence, enjoying each other's company in the gentle spring air, for what felt like hours. Though, in truth, it was little more than a quiet morning stolen by time itself. We watched the bees dance upon flowers, content in their ceaseless movement. It was a morning of silently bathing in spring as if the entire world had paused for our reflection.

Later, the day unfurled in its usual way. A light lunch, the promise of simplicity found in the bustle of an open market. We met Paul and wandered through galleries, Brit drawn to the art in a way that I, too, understood, but from a different vantage. There was passion in the brushstrokes, emotion hidden within the lines, for me; for her, it was the precision, the calculated yet spontaneous splatter of paint, the technical mastery. Paul, with his quick laughter and endless enthusiasm, offered explanations as we moved from one piece to the next, weaving through paintings, sculptures, and photographs. Each step felt like a quiet discovery, his voice shifting through accents, an unexpected joy in his words, as we moved from one piece to the next.

As evening crept upon us, the world softened into dusk, so we opted for an early dinner, a chance to rest before heading out for drinks.. Brit's idea of "the hair of the dog" had suggested it, and so we made our way to Soho, to the familiar restaurant that had long since shed its former dilapidation. Once a neglected Georgian townhouse, it had been transformed into a haven for those who craved creativity, where artists, musicians, and writers congregated. We had come many times before, not for the food alone but for the people, the atmosphere that buzzed with worldliness.

We were greeted by the barmen, who handed us glasses of Prosecco, but I, as is my wont, opted for a non, alcoholic spritz. My preference leans toward the clarity of presence, savouring the moment without the fog of drink. Still, occasionally, there is something to be said for a night of Irish cream ales. One of my colleagues, Emma, a curator with a thick London accent, also shares this indulgence. She turns the museum into a Disneyland of academia for anyone who meets her, big or small. On occasion, we make our way to a little pub, where we sip on creamy ales and discuss everything from Bronze Age artefacts to Ice Age relics, or wherever our conversations take us.

After a memorable dinner, with Brit starting to feel better, the night was drawing to a close. It had been a long day, having been up since 4 a.m., but it was one of those memorable days, long yet full to the brim. So, exhausted, I shuffled off to bed, content and reflective.

Chapter 4
One Male and the Orchard Line

Morning crept in, soft but insistent. I pulled myself up, shaking off the last threads of sleep. The crisp London air slipping through the window's gap, the Thames beckoned, restless, like something half-remembered, and I set off for a run, the hum of the city unfolding in the rhythm of my steps, my breath falling into step with the city's quiet stirrings. Back to the flat, a quick shower rinsed away the morning's exertion. Leaving me ready to wander through the treasures of the museum. But first, a detour.

Molly's shop was a snug greengrocer nestled in the Borough Markets, it beckoned like an old friend. Molly herself, an old schoolmate of my mother's, had been a steady presence in my life since I could remember.

I'd known Molly since, as she'd say with a grin, "*Rocks were soft.*" She had a way of taking the sharp edges off life, slipping me fruit and forbidden sweets when my mother's sharp blue eyes weren't looking. Over time, she became a source of wisdom and advice dished out with hearty humour, and a side of flirtatious mischief. Molly had a passion for life, banter and people, and her playful winks and cheeky remarks often set those around her at ease.

As I pushed open the shop door, the familiar aroma of fresh greens, sun-glazed fruit, and crusty bread greeted me. Molly was at the counter, mid-conversation with a customer, but the moment she spotted me, her face lit up in a radiant smile.

'How ya, my cutie!' she exclaimed, leaning across the counter to envelop me in one of her warm, no-nonsense hugs. 'Ah, Dee! Back to the museum grind, are ya?'

The accent between us rhythmic as we bantered, Molly seamlessly juggling her customers and their groceries as she always had. When the shop's morning rush began to ebb, we sank into easy conversation, catching up on the familiar rhythms of friends, family, and the ebb and flow of life.

As we chatted, Éireann appeared, a young friend of Molly's who often delivered goods to the shop. Her weary frame and downcast eyes spoke before she did. When she finally shared her troubles, her words spilt out like the unsteady pour of a too-full cup.

Éireann had been eking out a living in a rundown flat on the other side of town, her dreams of returning to Ireland and renovating a small place slipping further from reach. Longer hours, harder work, and mounting bills had left her spinning her wheels.

Molly's face darkened as her voice grew sharp with righteous vigour. 'I'm gutted, Dee,' she said, shaking her head. 'Look at her, bless. It's not just Éireann, it's everyone these days. My old neighbours can't even keep the heat on or put proper food on the table. Freezing under musty blankets, no money for medicine, it's heartbreaking. The middle ground is vanishing, swallowed whole, and the Uber-wealthy, relentless as tides, are ploughing the rest of us under, leaving only desperate workers thrashing to keep their heads above the murk.'

Her voice grew louder, the shop quieting as customers turned to tune in. 'Corporate slavery, of course, that's the truth of it, though they'll never say it out loud. All the talk, all the cleverness... it hides the grit beneath. Hard, endless graft for so little in return. Rents that climb like smoke, leaving people choking on their own hopes. And health care? A luxury, for those who can afford to stay well. And the crime wave? Just the beginning, really. People are turning desperate. At this rate, we'll be back churning butter by candlelight before long, bartering turnips, and fearing the plague again!'

The weight of her words hung in the air, but Molly, ever Molly, was quick to soften the moment. She turned to Éireann with a cheeky grin and a wink. 'Now, what you need, Éireann, is a good man with some delicious hindquarters to help lift those boxes for you!'

Éireann 's lips curved with a tentative smile, and a soft laugh escaped her. For a fleeting moment, her burdens seemed lighter. That was Molly's magic, a rare mix of fire, compassion and kindness, capable of turning despair into something a little easier to carry.

Éireann nodded, her shoulders easing under the weight of Molly's heartfelt words. There was a kind of solace in Molly's fervour, a vividness that held its own against the grey weight of the world. Time, I thought, seemed to be slipping backward, unravelling at its edges, not just in the frayed hems of material obsessions, but in the deeper, quieter undoing of the soul. How quickly the world seemed to conspire against hope.

Hearing Elizabeth's struggles stirred something in me. It was heartbreaking to see her doing it so tough, so I offered her a lifeline. 'Anytime you need a break,' I said, 'you're welcome at my place. You're a good soul, Elizabeth, and an honest grafter. I'd be glad to help.'

Her face softened, and her eyes grew glassy with gratitude. 'That I will, Dee, if it comes to it,' she said, her voice catching... 'I don't know what to say.' She stepped forward and gave me a firm, heartfelt hug.

As Elizabeth hefted another crate, Molly's energy flared back to life, a spark in her eyes, her lips twitching with some unspoken amusement as she slipped into motion. Hands deft, she reached beneath the counter and handed me a bag of apples, oranges, and sweet potatoes. 'Best in the shop today,' she declared with a knowing wink and a gentle nod. 'I stashed 'em... Just a feeling, girl.'

I dug into my bag, fishing out a ten-pound note, intent on paying for the wee stash she'd set aside. But Molly, ever unwavering, gave a small shake of her head, final as a closing door.. 'No, lass. I'm fine,' she said, her voice resolute but kind.. 'Take this money and do something good with it. Help someone who needs it.'

Arguing with Molly was a losing battle; her will was always as steady as her heart was vast. So I nodded, smiling faintly. 'Well, that I will, Molly.'

I thanked her for her kindness, gave her a tight hug, and then I stepped out into the day, the bag of fruit and greens swinging lightly at my side. The museum awaited its purpose calling, but my thoughts lingered in that tiny shop, a world within a world, where Molly's compassion defied the encroaching shadows of an unstable world put to rise.

When at last I arrived at the museum, Emma was already there, a familiar figure pacing lightly near the entrance.

'I thought you'd abandoned me. D,' she bleated... her voice bright, though her smile softened the reprimand. Her vivid copper-red hair was tied back in a ponytail, and a yellow pencil rested jauntily behind her ear, a cheerful splash of colour against the calm grey of the morning.

'Sorry,' I murmured, offering a sheepish grin. 'I stopped to chat with Molly and got carried away.'

'Ah, well, she does love a natter, doesn't she?' Emma replied warmly as we turned and entered the museum together.

Excitement wove itself beyond us, an unspoken current flickering beneath each step as we slipped along the hushed corridors. A peculiar vibe, something between mischief and purpose, hummed at the edges of this new venture, still half-formed but full of fun. This was not Curator's Corner.. This was something else entirely. Curator's Challenge.

A notion spun from idle chatter, now taking shape among the staff as a way to shake the dust from our usual routines. The premise was simple enough. Two curators, each armed with an artefact of their admiration, would champion its story, its singularity, contending and persuading, weaving their arguments like masterful storytellers. The audience, unseen yet pivotal, would watch, engage and pass judgment. A debate of relics, of whispers from the past, made electric, made now. By week's end, the curator with the most likes would move on to face another challenger in the next round.

We saw it as a way to invigorate history, to draw people in not through dry accounts but through our own enthusiasm, our love for the objects we'd chosen. Emma, with her deep knowledge of Early Medieval culture, had selected a Celtic sword... Its beauty, its mystery, etched with strange inscriptions, made it irresistible.

Emma herself was a force with her boundless energy and gift for making intricate details feel immediate. She didn't merely explain; she captivated. I knew, even before the first words were spoken, that she'd be hard to surpass. A rival, certainly, but more than that, a kindred spirit in this peculiar, shared passion for the past..

The museum was ours for now, the public yet to arrive. Usually, Curator's Corner was filmed against the polished stillness of the exhibition halls, but today we had chosen our offices. The informality felt fitting, a quiet retreat where ideas could spill out unbidden.

The sword lay before her, reverent in its stillness. She ran a careful finger along the blade, tracing its worn engravings. 'The inscription...' she murmured, voice low, as if the blade itself might respond. 'It's too faint to read. But it could be an incantation, for strength, for absolution... maybe even for divine aid.'

Something in her tone seemed to carry the weight of echoes, voices that had long since slipped into silence, yet still clung to the steel. Emma, as if drawn by their faint resonance, took a close-up picture, the lens capturing more than just the surface for later analysis.

Above the hilt, just below the guards, another marking. Faint, deliberate. *One Male.*

I leaned in, lifting my eyeglass, letting the dim light catch the lines. They were familiar. Almost too familiar. A pulse of recognition stirred in my chest. Then, it dawned on me.

'Emma,' I breathed, lifting my eyeglass for a clearer view. 'This engraving, is like the one on my penannular brooch. The similarity... uncanny.'

She glanced up, eyes sparking with something quick, something electric. 'Could be we've stumbled onto a puzzle,' she said, the words slipping out like a half-laugh, half-realisation.

A puzzle. The thought hung between us. The engravings, the markings, pieces of something larger? Or just coincidence? We weren't the type to believe in coincidences.

'Let's mull it over with an Irish ale later,' Emma said, checking the time. Just like that, the spell wavered, the present pulling us back. The cameras were ready.

'Sounds like a plan,' I said, but my mind was still on the sword. On the brooch. On the feeling that something, something important was just on the edges of our grasp.

And so, with the camera rolling, we began. Words tumbling calmly, her voice weaving history into something tangible, something real. She spoke as though she had lived these stories herself. I worked in quiet focus, each adjustment pulling me deeper, listening not just to her but to whatever the steel had to say.

When the episodes were released, the response was immediate. Comments flooded in, and whispers from across the digital void were drawn to the passion we had shared. And, as I had expected, Emma's enthusiasm, her irresistible charm, carried her ahead in the tally of likes. By the day's end, it was clear: her victory was all but certain.

As I sat in the quiet of my office, tidying loose ends, her voice caught my attention... 'My shout!' she called, leaning around the doorframe with a grin as wide as the morning horizon. It was a generous offer, a gesture of camaraderie from the undeniable victor, and I couldn't help but laugh.

Half an hour later, we found ourselves ambling to a little pub tucked away at the edge of a cobbled street, our favourite wee retreat, a place where time seemed to fold in on itself, wrapping us in an old school ambience. We slipped into our usual corner, and Emma, her smile both mischievous and inviting, declared, 'Well, my round then. 'Looks like we might sample a couple of cream ales, eh? It's been a while, but we'll endeavour to behave, D.'

Ah, the well-trodden road of good intentions, how often it takes a detour, unbidden, towards indulgence. Where one ale turned to two, two to three, and soon we were swaying, the world tilting ever so slightly. Yet even in our unsteady state, we retained a peculiar clarity, bound by a thread of curiosity to the task at hand, the mystery of the Celtic sword.

We began closely analysing the photo of the inscription: ND O ORVI. We leaned close, as if proximity to the words might unlock their meaning. Together, we pieced fragments of Latin, whispering, Nomini Domini Omni Orvi— "*In the name of Our Lord of all the earth.*" It seemed quite plausible, didn't it? And those two bold Xs etched deep, surely they symbolised Christ, or perhaps something more arcane. Not much, but there it was, the word like a sword, evoking something ancient, something intriguing... *hick.*

Emma had indulged in one Irish ale too many and had contracted a sudden case of the hiccups, but we stayed focused. The sword's true story, however, might have eluded us. Perhaps the engravings had been added centuries after the blade's forging, or perhaps they were little more than the whimsy of an idle hand. *Hick...* No matter. The pleasure lay not in solving the riddle but in the act of unravelling it, our laughter then mingling with the rich, frothy residue of our pints. Two friends entwined by history, by culture, by hiccups and by ale.

Emma, her cheeks flushed with mirth and light laughter, leaned back with a sudden recollection. '*Hick.* Do you remember,' she began, her voice lilting, 'the first time I dragged you here? You'd just started at the museum. God, you were scuttered. I had to bundle you into a cab, you left half your coat tail stuck in the door!'

I groaned as the dust shook off my memory, rolling my eyes, I replied. 'And the cab sped off while I stumbled behind, like a sozzled Gibbon at a ceilidh dance. Don't remind me.'

Hick, Emma, still burdened with hiccups, erupted into laughter, her head tilting back, eyes sparkling. 'Yes! And you were squealing, 'Stop! Stop! Stop! For feck's sake! Your handbag banging against the car door like a weighted tambourine...'

She wiped tears from her face, struggling to catch her breath. *Hick, hick.* 'I'll never forget the cabbie's face, utter disbelief, eyes wide, staring at you with your face pressed up against the rear window, smeared in lipstick, like a deranged zombie who'd just discovered its own reflection.'

Her bright, uncontainable laughter spilled over, and soon, I was gasping for air, caught in its infectious tide.

We tried, unsuccessfully, to steady ourselves with Emma's hicks and the under-the disapproving gazes of the regulars, their pints held firm, their eyes heavy with quiet judgment. It was time to leave...

The night air outside brushed against us, sharp, cool, its clarity startling after the warm haze of the pub. Something stirred in me, then a feeling I couldn't quite name. Tipsy as I was, a strange certainty lingered, as though the threads of the evening, of the sword's engravings, of my recent dreams that were weaving their way into something larger, something waiting just out of reach. An odd mix of déjà vu and epiphany.

When the cab pulled up outside Brit's, I wrapped my jacket around me with a quiet, deliberate finality, determined not to leave behind even the faintest trace of unresolved things. I waved Emma off, watching as her cab drifted into the distance, her bright, unshakable smile still glowing through the rear window, her hand lifting in a series of spirited goodbyes as she faded off into the night...

Still slightly wobbly, I began trekking up the stairs to Brit's apartment, each step creaking faintly beneath my unsteady heels. Inside, I set my bag down with a dull thud and made my way to the kitchen. Brit's wee coffee machine sat in its usual pride of place, a small beacon in the dimness, my small, glowing temple of late-night salvation. I jabbed at a few buttons, listening as it grumbled awake. The scent of coffee curled into the air, rich, reassuring, and grounding.

Cradling my mug, I shuffled to the doorway, wandered into thy room, nudging the door shut with my foot. With the gradual fading hum of traffic outside marking another day's end, I sat, exhaling, letting the evening finally catch up with me. Then, with a slow stretch, I flipped open my laptop, allowing the soft glow of the screen to draw me into my after-hours ritual of sifting through the world's latest murmurs, and any other subjects that came to mind.

Chapter 5
Whispers of the Dreaming

Last night's "Native American Dance" dream lingering in my subconscious, its imagery emerging occasionally, vivid and insistent. I decided to delve deeper, running searches on Native American dances. Sundance, Ghost Dance, and anything related to visions or dreams tied to these traditions. Then something caught my eye: a thread from the Wakan-Tanka Forum.

The name itself held weight. Wakan Tanka means "The Great Spirit" or "Great Mystery" in Lakota, Dakota, or Nakota. Often referred to as the Great Sioux Nation, it describes the sacred or divine power inherent in all living things. Wakan, meaning holy, and Tanka, meaning "great" or large, echo the reverence many across the Great Plains hold for the bison, known as Tatanka or "The Great buffalo."

I began reading through the forum sections, which were dedicated to various topics, all presented thoughtfully and with care. The site was meticulously managed, and its structure invited exploration...

There was a wealth of information on Native American culture and other Indigenous and European tribal traditions. But what drew me in most was the Vision/Dream section. It was filled with personal accounts, dozens of stories from people around the world describing heightened dream activity and their connection to prophecies or the Great Mystery.

The forum felt like a linguistic and cultural museum, a repository of traditions and perspectives that offered an intimate glimpse into ancestral knowledge. As I read, suddenly a strong feeling washed over me, a sense that these vivid dreams I'd been having had somehow led me here.

Though I was an academic at heart, one to cross the t's and dot the i's, my mind had always been open to the unknown. Mythology, the Great Mystery, the spaces between the rational and the mystical, had long fascinated me. The forum was neatly divided into categories: visions, dreams, prophecy, videos, and music. One section was entirely dedicated to historical information on Native American cultures, particularly those of the Great Plains.

As I browsed, I came across a thread that made me pause, a dream account eerily similar to my own Zeus dream, with its arm maze and undertone of something vast and ungraspable. The dream, titled "The Man on a White Horse," was authored by Anna "*White Hawk*."

I began reading:

I was lying in bed when a man on a white horse entered the room. His face and chest were painted with signs of the old ways, his arm marked with a spirit maze, like the paths walked before time had names. He did not speak, but his eyes held meaning. At first, I did not rise, but he beckoned, and so I followed.

When I stood, he began to shape objects with his hands. A great bone became a flower. Then another. And another. Everything he touched blossomed as if waking from stone. He offered me a heavy bone. I hesitated but felt encouraged by him to take it. The moment I did, it transformed into a flower in my hand.

My dream shifted... My clothes had changed into buckskins adorned with beads, herbs woven into the stitching, flowers at my chest. I stepped forward and entered a room where another man lay. His spirit was heavy, unwilling to rise. Behind me, many stood, white with light, filling the space with something neither seen nor spoken, yet felt. I sat beside the man, gently stroking his face. He saw the white ones too.

The room held shadows, thick in the corners, whispering. I pulled the man close, wrapped my robe around us both. The darkness then stirred, fading, retreating as mist before the sun.

The man on the white horse returned, lifted me onto his back, and we rode far. Across open land, until we came to the water's edge, where a great wooden boat sat waiting. Its sails, worn and strong, held the breath of spirit. People stood at its side, casting things into the waves, what they let go of, I could not see.

Then, we rode again. Another man lay still, his body blackened, burned like wood after fire. I took a breath, deep as the roots of the earth, and exhaled. The air carried life into him, and the charred places softened, healed. Pain left him.

The white ones moved to a woman, her body nearing death. They gathered around her, hands outstretched, their light weaving itself into her, as a river takes in the rain. And she was restored.

Anna

Though Anna's dream was not identical to mine, there was a shared quality, a thread connecting our experiences, the man, the symbols, the sense of something profound yet just out of reach.

I leaned back, letting her story settle, its imagery vivid and mysterious. But the maze caught my attention the most. This didn't feel random, not some passing alignment of thoughts and timing. It had the weight of something intentional, like a door easing open just enough to glimpse what lay beyond. An invitation, maybe. Or a pull toward something I couldn't yet name.

I wanted to understand more, not just the surface of her thoughts but the shape of the mind that held them intricately. So, I shared my own dream, the one about Zeus, on the forum. Not just to put it out there, but in hope she would respond, that she would read between the lines the way she seemed to do so effortlessly with others.

Anna responded with a thoughtful reply. She believed her dream was tied to an old story about the spirit of Wakan-Tanka taking human form, or connecting with a human form, to guide us, to reveal the power within. There was a grace in her wisdom, an openness that illuminated the beauty of Native American culture, rooted in harmony with nature and a shared destiny with all living things.

In her dream, the great spirit had taken the shape of a man riding a white horse, teaching her about transformation and sacred potential. She told me the dream was about connection. Not just to people, but to everything, the land, the slow pull of unseen forces...

Then came the Zeus dream. The snake, she didn't think of it as just a symbol. It was a path unfolding, a shift, something old stirring, awakening. Her words sat with me, like the hum of something half-remembered, and I felt it too, that shift, deep, something ancient stirring awake, but somehow new.

It brought me back to when I first read about the great Native American chiefs. They weren't just warriors, or visionaries, or leaders. They were stillness, the kind of pillars that hold everything together. One of my favourites was Luther Standing Bear, author, educator, and philosopher. He expressed words that stayed. Something about the way he spoke, not just of his people, but of the space between worlds. He didn't just hold history; he threaded it together...

And as I browsed the Wakan-Tanka forum, one of his quotes came to mind:

"*Despite all the teachings and doctrines of mankind's great religions, humanity remains still confronted with the Great Mystery*."—Luther Standing Bear.

I wandered through more dream stories on the forum- each one drawing me in with its intrigue and quiet depth. It was an unexpected revelation, a window into tribal cultures through voices that spoke from personal experience, far more intimate than anything I had encountered before.

There were spiritual beliefs, too, delicate threads I hadn't come across in my research on cultural artefacts and their symbolism. What lingered most were the interpretations of dream symbols, so vivid and unique, woven with ancestral stories and prophecies handed down like whispered secrets through generations. These voices, their stories seemed to unlock something larger, layering dreams, culture, and the connective tissue of shared human experience with a quiet, profound richness.

At that hour, weariness draped itself over me, a quiet lethargy brought on by the long day's toil and the gentle lull of a few pints. I decided to log off the computer and give Mum a quick call to check in on how Dad, wee Puddles, and the rest of the family were doing during this difficult time. I tapped Mum's number on my mobile and hit the video icon.

Her face appeared, warmly lit and framed by the kitchen behind her. 'Hi, Mum!' I said, a smile rising despite my fatigue.

'Hi, darling! How are you doing, my girl...?' she asked, juggling the phone in her hands.

'I'm good, Mum. Just checking in to see how everyone's getting on,' I replied.

She smiled, radiant in her way. 'We're all doing okay, blossom. Your father's holding up, he's been out and about in the spring air.' Then, Puddles emerged over her shoulder, his face appearing cartoonishly expressive as ever. The sight made me laugh...

Mum caught the moment, her voice lifting. 'Now, what have you been up to, eh?'

Still laughing at the trace of Puddles lingering, I said, 'Researching, Mum. Always researching. And winding down after a creamy ale or two with Emma.'

I hesitated, then decided to share. 'I had one of those clear, powerful dreams again, Mum. So I started digging into it, and came across an interesting forum...'

Her face softened, eyes bright. 'Oh, that sounds grand..! Family tradition, those vision-type dreams of ours. It's good you found a place to share.'

How right she was. On Mum's side, especially, our family seemed to carry this thread of profound dreaming, a gift, some might call it. My grandmother had always spoken of the "Old Book," her term for Scripture. To her, it wasn't a rigid text, but a tapestry of dreams and visions, metaphors for the unknown forces that touch people's lives.

Mum's words carried me backward into the folds of memory, where my grandmother's stories lived. Perhaps it was these series of dreams unfolding that rekindled a vision she'd described when I was born, tied to the Scripture that proclaimed: "*In the last days, God will pour out His Spirit on all people. Your sons and daughters will prophesy, your young will see visions, and your old will dream dreams...*" Curious, I thought, that this legacy of dreams would linger with me now, despite my own leanings toward agnosticism. Yet, as a curator, my work often intersected with these realms. So many artefacts were steeped in mythology or born from humanity's yearning for the divine, for meaning beyond the tangible. Puddles interrupted my thoughts in the only way he could: pressing his damp nose into the screen, demanding to be seen. I laughed again, nudging my own nose toward the camera in a gesture of affection, cooing those nonsensical "miss you" sounds that Puddles adored.

Mum's laughter joined mine, lifting the moment before our conversation turned again to dreams.

Her eyes brightened, and she leaned closer. 'I had a dream as well the other night, Dee, a right strong one. I dreamt there was a huge tree growing in the middle of your house. Its branches were so big they burst through the roof, opening it to the sky. The blue was so vivid, yet it was circled by these swirling dark clouds. I was so scared for you. But then, this amazing blue light shone down from the sky, filling your whole cottage, with calm. I could feel it and see everything, so clearly.'

She paused, her voice softening as she gave her interpretation. 'I felt the tree was the tree of life, yes, and that light... it emanated with renewal.'

Her words, vivid and full of meaning, left me with a sense of connection, somehow to the sequence of dreams unfolding.

I hung up, the conversation lingering in the air, fulfillment settled over me, as though I had stumbled upon a long-sought book, its pages now unfolding before me. The day had unfolded in a way that felt... deliberate, moments slotting into the next, following some unseen thread. Mum's dream serving as the perfect conclusion, on an already intriguing day.

Donning my dressing gown, settling onto Brit's deck. The spring London night air greeted me, whispering something intangible yet comforting. I couldn't quite grasp its message, but it soothed me nonetheless. Then, carrying that stillness with me, I switched off the light and let sleep take me, deep and dream-heavy, like slipping into some vast and waiting sky..

The following morning, I awoke to the sound of birds chirping, their wee melody interrupted by the rude cough of a moped's back-fire. It was Saturday, and my head felt glued to the pillow until a loud rat-a-tat-tat on the door unstuck it. Only one person knocked like that. Paul. With that, I knew the weekend had officially begun..

I opened the door, and there he stood, a precarious collage of art supplies and an easel jostling in his arms. His grin was infectious as he declared his mission: he intended to immortalise Brit and me on the balcony. But Brit was out, off on her Saturday ritual, returning with eggs and coffee beans, so we waited.

When she returned, she seamlessly slipped into her quiet breakfast ritual. I toyed with the idea of sharing the peculiar events of the previous day, including my encounter with Anna "*White Hawk*", but decided against it. Today was meant for being in the now, for settling into the unhurried cadence of Brit and Paul's company.

We talked and ate, poached eggs, toast, coffee, moving easily from one conversation to the next, until we drifted outside to the balcony, where Paul set to work. He launched into it with a feverish energy, white charcoal darting across the canvas as he narrated the week's chaos and creative triumphs. After a while, he disappeared momentarily and returned with a giant glass of red wine in hand.

Brit arched an eyebrow, saying. 'Renoir, you'll be sozzled by midday! Isn't it a bit early?' Paul struck a mock French pose and declared in a dramatic accent, 'Madam Snookums, for art par excellence, one needs a big dollop of red wine, oui!'

'Now, Dee,' he continued, squinting while measuring with his charcoal, 'hold that cup to your lip. Don't move for five minutes!'

I realised we were in for a proper still-life workout. My wrist started cramping, while Brit, legs propped on the balcony rail, muttered about pins and needles in her feet. We relaxed when Paul transitioned to his paints, though we tried to keep somewhat still as he dabbed and swirled colours with intent, coaxed into artistic alchemy.

Hours passed, and finally, he stepped back, surveying his work with a flourish. 'Done!' he declared...

Paul's signature style leaned toward expressionism, but to my surprise, he had captured us in impressionism. The painting bathed Brit and me in soft, warm light, surrounded by her potted and cascading flowers, framing us perfectly, the lush, vibrant colours were almost alive on the canvas.

'Oh, Paul,' I murmured, 'awe blooming, this is mine. For my cottage wall. I'm buying it.'

He smiled with a paintbrush dangling from the corner of his mouth, his pride evident, but Brit's eyes sparkled with mischief. I recognised that look, it meant she was ready to wrestle me for it.

Paul leaned against the rail, affecting a practised nonchalance. 'I must say, Ms. Dee, I've truly outdone myself.' With a flourish, he extended the painting toward me... 'It's yours, a freebie.' I wouldn't hear of it. Knowing Paul had been going through a tough time and that this piece deserved far more, I insisted on paying. Thinking of Molly, I calculated quickly, 'Kindness is a virtue,' and handed over £200, a bargain for something so personal and beautiful. Paul protested but quickly realised I'd made up my mind.

Brit just smiled, flicking my hair back playfully. 'Well, I have to let you have this one, eh?' she teased, planting a quick kiss on my forehead.

We ended the afternoon with Brit's homemade pizza, and Paul taking delight in theatrically tossing the dough. That evening, my film choice fell to Whale Rider, a quiet reflection of mythology that lingered in the recesses of my mind.

After the movie, Brit and I dried our eyes, and we said our nighty nights. Back in my room, curiosity nudged me toward the forum. There were no new posts from Anna, so I added my own, recounting the dream in vivid detail, the ghost dance, the enigmatic woman, and the feeling of connection it stirred. And, as always, I tumbled into research, delving deeper into the stories of the past, ghost dances and untold histories, before sleep drew me under once more.

Chapter 6

Lines Between Worlds

Somewhere between dream and daylight, the kitchen called. A good cup of coffee answered. Then, an odd feeling came to me, so I checked the forum. Logging on, I saw that Anna had responded to my post about the dream and the dance, and I eagerly began reading her reply...

Anna wrote:

Firstly, D,

My mother and I walk the path of strong dreamers. Dreams are sacred, not merely shadows of the mind but threads woven between this world and the spirit world. We know that what the eyes see is not all there is, beneath the surface, the unseen moves, shaping what is and what will be.

When I was young, I told my mother about a dream, one that carried weight, the kind that lingers in the bones. Not long after, a holy man came to my father and said, "*Your daughter has dreamed.*" He spoke of the Creator's gift and of a choice that was mine to make. I listened, and I chose to dance, the Sun Dance.

The holy man walked with us, guiding my family in the ways of preparation. The Sun Dance is not a thing one steps into lightly it is a commitment, a prayer carried in the body, in the breath. A year of ceremony, of cleansing, of learning. And then, at last, we stood upon the sacred ground of White Buffalo at Green Grass.

I danced. I gave all that was in me, until I could give no more. My body faltered, but the spirit carried me. And in that place, in that moment between exhaustion and awakening, a vision came to me, so powerful it filled my chest until I wept, not from pain, but from the vastness of it, from the knowing. There are no words to shape such things, D, not fully. Only gratitude. *Thank you, Tunkashila. Thank you, Great Spirit.*

Tell me, D, the dancer in your dream, what did she wear? What were her eyes like when she looked at you? You said she showed you an old picture, yet she had not aged. There is meaning in this. Look again. What else do you remember?

Blessings,

Anna

After reading Anna's reply, I felt deeply moved, and a strong, intuitive feeling came over me. I recalled the dream in more detail and responded:

The dancer, Anna, she was dainty, yet was strong and determined in her demeanour, especially in the way she repeatedly showed me the correct way to do the dance. I also remember she had something wrapped around her wrist, a shawl, I believe.

The shawl was naturally dyed and had a symbol on it, perhaps half of an eagle. I vaguely remember it had two or three feathers attached, white feathers with black tips that stood out. Another detail just popped into my mind: I'm quite certain the shawl was mainly blue.

Anna, thank you so much for sharing your story. I feel the heartfelt importance of it all. I'm aware of the Sun Dance and Ghost Dance, as my profession is that of a curator. I've studied many Ghost Dance dresses and have done a lot of research on your incredibly rich culture. It's been both a personal passion and something of great historical value to humanity.

I understand how the Sun Dance, over many generations, has served as a bridge to the spirit world and a guiding force in the natural world.

Kind regards,

D

After I had replied, I thought it was time for a bowl of fresh fruit and oats, with a good splash of almond milk, followed by my ritual chill session out on the deck. An hour or so later, I checked back on the forum.

Anna had replied:

Oh, D... my heart is full, and my eyes are wet with tears. That blue shawl, my grandmother's hands wove it with love, and I hold it now as I write to you. There is a white hawk upon it, just as you saw. The feathers, too, are white hawks, carried through generations, passed down from our ancestors.

I am older now, but the way you saw me... it is how I was then, and in truth, how I remain. The fire within me has not dimmed. I have walked this path with steady feet, learning the old ways, holding them close, keeping faith with all that was given to me.

D, I believe the dance was in spirit, and you stepped into it with such grace, just as you have stepped into this moment, finding me here, across time and distance. This place, this gathering of words... it is sacred in its own way. Thank you, Tunkashila.

Blessings to you,

Anna

I must admit, reading Anna's reply brought me to tears. It was as though something inside me had been released.

I promptly replied:

I feel that too, Anna. I really do. The dancer in my dream was you, it all matches up in detail. I'm a bit overcome right now. I made the right steps, thanks to you, and I'm so glad to have found both you and this forum.

I feel there's something strong at play here, and I must say, while I'm usually a very grounded and meticulous person, at the moment, I'm feeling quite floaty, as though I'm on some mystic high.

Blessings to you, too, dear Anna,

D

Anna replied via a DM:

D, we will switch to email, as there are things best spoken there. For now, I have errands to tend to, family matters that call me, but we will speak again soon.

Also, if you choose to post on the Lakota thread, do not be troubled if two women challenge you. I have spoken to them once already for their lack of respect, but I feel their hearts are not in a good place. They may come against you, not for truth, but for the weight of your knowing. Do not let it shake you. I will be watching.

Walk well, and we will speak again soon.

Blessings,

Anna

To which I replied:

Thank you, Anna. I'd planned to contribute to a couple of threads, but if anything untoward occurs, I'll refrain from letting my feisty Irish side create too many sparks. lol.

Love,

D

So afterwards, browsing through the other threads. One post began to take shape in my mind, centred on Lakota tepee etiquette, a subject I'd stumbled across and found both intriguing and quietly profound. The idea felt close to home, linked by memory. My thoughts drifted back to Australia, to a time spent on a windswept seaside property. There, a tepee had stood, humble and inviting, fashioned as a guest dwelling. Nights spent within it felt both majestic and grounding, the distant crash of waves rising and falling like a heartbeat as I drifted to sleep. A vivid, soothing memory now returned.

Another thread followed, one steeped in visual echoes of the past. I shared images by Edward S. Curtis and Frank Rinehart, their deep photographs, strong and quiet as prayers... Images of Native American chiefs, leaders, and communities gazed back, their dignity etched into each frame. The strength of those images carried something rare, a presence almost forgotten to us in today's world. They held a resonance, a timelessness that lingered. I shared what I had, carefully pairing each image with notes, details, and histories, hoping to honour their depth and invite reflection.

When I had finished, the morning felt lighter, yet my joints carried a faint whisper of wear. The idea of an exercise bike, prosaic but practical, had been circling my mind for days. My outdoor juants, once my ritual, against rain and wind, had over time tugged at me differently, as though asking to be reconsidered. The cold mornings had grown less thrilling, the pounding on pavements a persistent reminder of limits. Wisdom, I thought, was in adaptation: the choice to preserve rather than overstrain, to sustain rather than resist. An occasional park run in spring, I vowed, would remain, soft grass and warm air still held their place, but the day-to-day called for a change to preserve my joints.

I thought of the local fitness shop; the people there were always genuine and informed. I preferred to support them if I could. The shop felt familiar, its rows of machines and other fitness equipment quiet and waiting, its staff quick to assist with practical advice, never intrusive.

The bike I'd envisioned was there, its lines simple yet purposeful. The decision came easily, as though it had already been made: I would take two, one for myself in London and one for Mum, to be collected on my next visit to the cottage. It felt good to invest in something both practical and beneficial. Puddles, no doubt amusingly, would be intrigued by the sight of me barrelling along indoors.

I wrestled one box from the shop, hauled it out of the cab, and wriggled it up the steps. I thought I'd leave this to Brit to assemble; she loves getting hands-on and studying manuals. No doubt, when she got home, she would dive right in, probably logging a few kilometres to break it in.

I then checked back on the forum. Some great comments and appreciation had been posted, as I had made a big effort to research and include all the details, especially with the wonderful old plate photos. However, in the comments under the Tipi Etiquette thread, someone, presumably of Lakota origin, had written:

"You can't post things like that. These are things only Native Americans can inform people of, so it's best you remove your post!"

To which I replied: "This is well-documented information, and etiquette like this is available to anyone. It's not a sacred ceremony reserved exclusively for Native Americans; it's respectful etiquette. Etiquette is something people research when visiting different countries and cultures. It's a mindful and respectful thing to learn and share."

Then another female poster chimed in, saying: "I bet she has dream catchers hanging up everywhere," adding a laughing emoji to top off her condescending remark.

I also noticed a comment from one of the same women on the Historical Photos thread: "Posting things like that makes people think we all still live in tipis!"

I responded: "These are historically honourable images of great chiefs and respected men and women. Posting them is purely out of reverence and respect for the great ancestry of the Plains people. And who in the world would think Native Americans all still live in tipis? Not that there's anything wrong with living in a tipi, it's a wonderful and functional space regardless!"

By then, my Irish blood was up. I thought to myself, okay, D, take a breath, step outside, and chill for a moment before saying something you'll regret.

When I returned, ready to defend myself again, I noticed that all the unsettling comments had been removed!

Anna had sent me a private DM:

D, I offer my apologies. Prejudice knows no boundaries, it is not held by one people alone. I have seen this before. I had hoped their hearts carried more depth, so I gave them the chance to walk rightly, but their ways were not good. Their words were not of respect, and so they are no longer here. Your voice was steady, your words honourable. They were heard and valued.

Blessings,

Anna.

I replied, with a note of understanding:

It's alright, Anna. I admit, I was a bit heated because my intention was solely to be respectful and informative. I do recognise, however, that some Native Americans might feel uneasy when individuals of European descent share or discuss their culture in certain contexts.

Kind regards, D

Anna responded:

No, D, you must not make peace with what cannot stand. They cannot wield the name of our people as a weapon to silence others. I have seen this before, those who know little often take offence when their own ignorance is laid bare.

John Trudell spoke of this with a clear and fearless heart. He was a warrior of words, a poet who carried truth like fire. His voice still echoes, D, look for - Being Human. In it, he speaks of how some fight only to guard the label of being "Indian," when that word itself was never ours.

We are not "Native Americans." We are not "Indians." We are something far older, something no single name can contain. We are the People. We are human beings, born of the land. Respect and heart, these are what matter, D. If we, who have carried the weight of history, forget them, then we walk the same cold path that was once forced upon us. Mistakes can be mended, but when a heart is filled with bitterness and pride, it poisons the spirit. That, I can not allow here.

Blessings,

Anna

I replied:

I stumbled upon John's videos a few years ago, and they have been a quiet current running through my life ever since, steady, unwavering, always there in the background. His voice, both fierce and tender, has clarified the way I see the world and the way I listen. His poetry doesn't just live on the page; it lingers.

That video you mentioned. I haven't seen it yet, but I will. He saw with a rare kind of clarity, the kind that leaves an imprint long after the words have been spoken.

John belongs in the same breath as Martin Luther King Jr., a visionary, a truth-teller, a force of nature. It still baffles me that his name isn't spoken more often, that his brilliance isn't as widely known as it should be. I do what I can to change that, to pass his words forward in the hope that they might land where they're needed most.

Sending love and peace your way, D

Anna's calm authority and steady wisdom created a sense of quiet assurance, leaving me free to explore the forum at my leisure. I moved quietly through forum sections, lingering, reading, and taking in the vibes. Now and then, I left a comment, sometimes starting a thread of my own. One in particular, was on Ghost Dance dresses and shirts, seemed to spark something. Others joined in, sharing their thoughts, appreciation, and their quiet reverence for the history woven into those garments. There was a kind of unspoken understanding between us all, a recognition of something deeper, stories stitched into fabric, of memory, a hope held in every thread.

The rhythm of the day shifted. I heard Brit return home, her familiar chaos of energy filling the space. From the sound of it, she had already begun tearing into the exercise bike box I'd messaged her about earlier.

Drawn by the commotion, I wandered into the living area and was greeted by the sight of Brit, locked in a glorious battle with cardboard and bubble wrap..

'Toolbox in the laundry, D. Fetch it..!' she commanded, eyes flicking through the newly unearthed manual as though deciphering ancient scrolls. A pause, a sideways glance. 'And a cup of tea. No, make that a mug.'

'Yes, ma'am,' I murmured, gliding past the kettle's click and hiss already in motion as I headed for the laundry. Moments later, toolbox in hand, I returned, setting down her tea, a steaming, earthy offering.

A quick succession of instructions followed - 'Hold this,' 'Grab that', each thingy clipped in and precise. Under Brit's focused hands, the bike took shape with formality, like a puzzle snapping into place.

'All right,' she said, wiping her palms with satisfaction. 'Time to give this puppy a spin. Grab me a water bottle..!' She swung onto the saddle and began pedalling, a look of fierce purpose stealing over her face. She adjusted knobs, pressed buttons, with each movement a subtle calibration of ambition.

I leaned back with amusement curling in my chest, watching her pedal furiously in her bib and braces, looking every bit like a Tour de France contender. She must have hit a hill climb as she lifted out of the saddle, her hair swaying rhythmically. Taking a drink from her water bottle, she sprayed a theatrical squirt down her face, then launched the bottle aside with vigour, where it bounced off the couch and went skittling across the floor.

'Another bottle, D!' she panted. 'Quick, like..!' - leaving me in stitches.

After a while, she seemed to hit the 200-metre mark, low in the saddle, going full throttle with a whizzing sound intensifying. Then, a breathless declaration: 'I'm going for the Green Jersey, D!' Her words triggered me into breathless laughter.

At last, with a final thrust, she lunged forward, the bike's base emitting a triumphant squeak against the floorboards. Her arms shot up, fully extended skyward in victory. 'Eat my shorts, Cavendish..!' she crowed.

By then, I was in tears, collapsing onto the couch, utterly undone by Brit's stage win.

Exhausted, Brit dismounted the bike, her legs moving stiffly as she made her way towards the shower room, a road warrior retreating from battle. My turn. I eased onto the saddle and took the bike for a whirl, not an epic climb through peaks and valleys, but a smooth, unhurried ride down a rosy country lane. Watching Brit wrestle the bike into existence meant that when I picked up my own in Ireland, assembly would be a breeze. I felt a quiet satisfaction in my purchase as I peddled onward, the relief of avoiding torrid weather that had now become more of a weary burden than an exhilarating challenge. The bike would also offer a gentler reprieve for my joints.

Paul arrived home, eyes narrowing in mock appraisal... 'Spunky set of wheels, D. You cardio queen, you.' With the remnants of Brit's packaging chaos strewn about, we gathered up the debris and ferried it out to the recycling bin.

Evening descended like a comfortable blanket. Brit and I moved through the kitchen, the rhythm of chopping and stirring falling into sync. Paul sprawled on the couch, languid as a sun, drunk cat, the remote control a sceptre in his hand, clicking through channels with casual indifference.

Evening descended like a comfortable blanket. Brit and I moved through the kitchen, the rhythm of chopping and stirring falling into sync. Paul sprawled on the couch, languid as a sun-drunk cat, the remote control a loose sceptre in his hand, clicking through channels with casual indifference.

The news flickered on. Paul grimaced, his voice a theatrical groan. 'Oh gawd. Electric Jesus's Grinch of Tangerine and their loyal lickspittle are at it again, my Brit.'

Brit's knife continued its steady rhythm against the cutting board. She replied with serene conviction. 'To paraphrase an old proverb, Paul': "*The forest was shrinking, but the trees kept voting for the axe; for the axe was clever and convinced the trees that, because his handle was made of wood, he was one of them.*"

Paul sighed; a sound laden with the weight of inevitability. He flicked to another channel. 'True enough, Brit. But at this rate, as the world hurtles towards Hades, the man with his quartet of steeds will arrive, brimming with wrath and fully charged.'

'Well, Paul,' Brit said, glancing over her shoulder, 'leap upon your steed and make haste to the shop, we're out of Parmigiano.'

Then Brit brandished her wooden spoon... thrusting it towards me with the flair of a paparazzo's microphone, a swift, precise motion that froze me mid-reach at the fridge door. Her eyes held steady to mine with faux seriousness. 'So, D, what's your comments on the current political forecast over on- Planet America..?' she asked, her voice a measured monotone, the kind news reporters use when they're fishing for a hit piece.

I lingered in the fridge's chill, a momentary refuge, before leaning toward the wooden spoon. 'Well, the far-right tribe thrives on intense ignorance, while the far-left tribe peddles a polished arsenal of passive aggression. Two wings though, same bird, Ms Brit,' I said, meeting her exaggerated reporter expression with a knowing glance.

'And that bird, Ms D?' Brit's tone took on a sharper edge.

'The Eagle of strength, freedom and peace,' I answered. 'It's sadly been descending into a vulture upon the poor and downtrodden for some time now.'

I straightened, shutting the fridge door with finality, as Brit tilted her head, eyes locked onto mine. 'Any last words, Ms D?'

'Behold, Ms Brit, the grand spectacle of Corporate America now comes complete with its own fully endorsed gazebo, proudly on display at the grand White House Hotel.'

Brit raised an eyebrow. 'So, will there be bouncy castles on the White House Hotel's lawns, seeing as it's become a layover for corporate clowns and crypto jesters?' She arched one eyebrow with added intrigue. 'Word is on the street, the swamp wasn't drained, it only got topped up with premium-grade sewage. Thoughts?'

'No comment,' I said, leaving Brit and her spoon behind, hustling my way out of the interview toward the lounge with a lovely bowl of tagliatelle nosh.

Brit's sister Sapphire arrived just after dinner, slipping through the doorway like a mystical ballerina. Her given name was Elizabeth, but a mystic once glimpsed the shimmer of Sapphire in her aura, and ever since, she insisted that everyone honour her ethereal self.

She glided in, eyes agleam with a softness that spoke of purpose, reserved for someone on the cusp of revelation. The plates lay cluttered on the table; Paul lounged on the couch, one hand hanging lazily over the edge, with the other scrolling half-heartedly through his phone. Brit leaned against the kitchen counter, arms folded, a glass of wine dangling from her fingertips.

Sapphire's eyes widened with the intensity of a newbie stage performer before she delivered her opening line. 'I have news,' she declared, her voice a measured melody.

Brit, ever so casually, raised her eyebrows and tilted her head. 'Let me guess, you're carrying, and plan to name the child Mud Brick.'

'Ha ha, sister,' Sapphire retorted, with a sarcastic smile. She was used to Brit's jabs by now. 'Well, I'm going to a mountain retreat,' she announced, her hands making slow, deliberate motions, as though she were conjuring up a temple into the room. 'I feel illuminated just thinking about it.'

Brit snorted delicately into her glass. 'Another self, realisation trip? You remember the fasting one, don't you? The three exquisite meals a day, each featuring a single lentil. You woke up in hospital, self, realised as dim-witted Sis.'

Sapphire took a breath, the serene expression faltering for a moment before recalibrating herself. 'Well, we all must embrace our journey and find positivity from our past learnings.'

Paul's voice oozed lazily from the sofa, his eyes fixed on his screen. 'Tell me, dear Sapphire, how many quid for them to reveal "the *Secret*"?'

Sapphire's brows twitched. She hadn't expected Paul to expose the retreat's thinly veiled commercialism so effortlessly, perhaps of enlightenment, for sale. But with a practised flick of her hair, she smoothed away the crack in her composure. 'It's not about money, Paul. It's about transformation.'

Brit rolled her eyes so far back, she could see yesterday. 'It's all drivel, Sap. Religion minus the bard. They take your money, yes, but at least you're allowed to leave without the threat of eternal damnation...'

A shadow crossed Sapphire's face, the warmth thinning to a chill. 'I'm sensing negative energy,' she murmured, her voice light yet edged with frost.

Brit sighed, setting her glass on the counter. Before Sapphire could flounce out completely, Brit crossed the room and pulled her into a firm, reluctant hug... 'Alright, I love you, Sis. Sometimes, it's tough love, but it's still love. Send me selfies at sunset so I can keep track of you.'

Sapphire's shoulders softened. She gave hugs all around, a delicate brush of affection to Paul, a slightly longer squeeze to me, before slipping away, her poise as ethereal in departure as it had been in arrival.

The door clicked shut behind her. Brit exhaled, shaking her head with a resigned smile. 'She's going to levitate one of these days.'

Paul quipped under his breath, 'Not on a diet of lentils.'

And the room, once again, settled back into its ordinary rhythm.

Chapter 7
When the World Tilts

After three weeks into London life, the city had started to seep into my bones - its rhythm, both frenetic and fluid, a strange paradox I was beginning to appreciate. The peaceful, lengthy halls of the museum didn't keep me still, my hours there lengthened due to the careful requirements of curation. By the time I pulled myself away from long workdays, the dusky twilight always hung low, and my legs felt like they'd been forged in the same era as the fossils I handled.

But there was a counterbalance to my grind, some things that made the weight of the days easier to bear. A community gently humming just below the surface of the world- Wakan-Tanka. A forum rooted in earthly wisdom, where digging deep into soil, into history, was more than metaphor. Not just pixels and text, but a web of voices, each thread woven with intent. Conversations spiralled: permaculture secrets passed like old incantations, self-sufficiency hacks that felt like forgotten lore, musings on ancestry that stirred something primal. Across continents, strangers planted ideas, and somehow, we were all tending the same unseen garden.

It felt good to drop some digital anchor time there, to circle through conversations that mattered. I learned. I contributed. I let myself be a part of this gentle ecosystem of grounded, earthy minds. There was comfort in knowing that while old London bustled around me, this quiet, steadying force thrived in the background. It tethered me to something more elemental, more real.

Brit and Paul, too, became anchors in this swirling new reality as our kinship deepened. Brit's flat was a quiet harbour, a place where the sharp edges of city life softened. The three of us fell into unspoken rhythms between camaraderie and comfort. We rode the crests of conversation and dissolved into laughter that sprang up unpredictably, like geysers in a bubbling landscape. Brit's dry wit cut through the air like a blade, always landing with that knowing tilt of her head, her mind sparkling with mischief. Paul, our self-appointed maestro of couch-surfing, the lounging philosopher, dispensed his quips like a languid oracle.

In the quiet moments, there was a contentment that hummed between us. Brit, chopping vegetables with the focus of a surgeon, while I pedalled away on my exercise bike, the rhythmic whirring filling the spaces between our breaths. Paul sprawled on his usual throne, sketching, or flicking through news channels with a performative groan or two. The mundane became ritual; the ordinary, somehow sacred.

Evenings were a blend of stillness and story. Plates of food passed between us like offerings. Tagliatelle slick with sauce, bowls of fragrant stews, bread torn and shared. The background hum of the television faded in and out as we traded anecdotes. The museum. The forum. Threads of digital connection intertwining with the tangible threads of common ground. Brit would occasionally thrust her wooden spoon at me with mock seriousness, demanding my "expert opinion" on everything from politics to the consistency of the sauce.

At night, when the flat settled into silence and the city's heartbeat dimmed to a murmur. I sometimes found myself back on the Wakan-Tanka forum, the glow of the screen and balmy nights softened my ponderings. The words of strangers, no, not strangers, kindreds, offered a warmth that felt like a blanket draped over my shoulders. We were scattered but connected. Different, but aligned.

I phoned Mum regularly... keeping her updated and checking in on everyone's well-being as Puddles scanned me, lurking over her shoulder. With Dad, our conversations often meandered through stories of my latest antique market finds, each peculiar discovery sparking his curiosity and drawing us into kindred discussions. I shared glimpses of the museum's quirks, of its flat's offbeat dynamics, painting a picture of my days with both warmth and humour.

The forum and Anna had become an oasis of sorts. I contributed my knowledge of history, swapping insights with people who spoke in the language of roots and resilience. In return, I received tales of homesteads in Montana, rooftop gardens flourishing in Seoul, and family traditions revived in the quietest corners of the world. It felt like being part of a secret council, a collective whisper of a future not yet lost, one firmly grounded in connection and renewal.

London hadn't dulled me as it once did. This time, it had sharpened and clarified me. Friendships anchored me, community steadied me, and the quiet thrill of possibility carried me forward. Maybe, just maybe, everything was coming together, albeit in the most unexpected and beautifully unpredictable ways.

I took a break from the museum after Paul rang. His voice was buoyant, brimming with enthusiasm, and he asked me to meet him for lunch at a cafe tucked just off the main road. The place was a charming jumble of mismatched chairs, peculiar artwork scattered across the walls like half-remembered dreams. I slipped inside, welcomed by the heady scent of freshly brewed coffee, the promise of something flavoursome, and the indistinct murmur of contented chatter. I chose a window seat, where the light softened the edges of the day, and waited for Paul.

A few minutes later, Paul whizzed up on his sleek electric bike, his pride and joy, his way of flipping the bird at London's relentless traffic and predatory parking meters. It suited him so well: efficient, practical, but just stylish enough as he had added personal touches to hint at the man behind the careful frugality. The British weather, often in a downpour, disagreed with his decision, but he never yielded.

He breezed in, helmet tucked under his arm, grin wide and broadening on his face. 'Alright, D?' he said, collapsing into the chair opposite me. Before I could respond, he spilled his news, words tumbling out with reckless speed, his voice bubbling over with enthusiasm. His artwork was going to be exhibited, a real, major showcase, and the exhilaration in his voice made him seem weightless.

I nodded and smiled; his excitement infectious, as we then nattered away. But something tugged at me just on the edge of my vision. Outside, two figures hovered near Paul's bike. Hoodies up, one with a backpack slung low. Their eyes shifting over the bike in that way, that sets off silent alarms, a tension that curls up the spine that whispers of trouble...

'Paul,' I whispered, my voice low yet sharp. 'Eh, Something's up.'

He followed my gaze. The light in his eyes dimmed, replaced by something sharper. Without hesitation, he was on his feet, striding for the door. I followed my pulse quickening, watching one of the figures pull a pair of bolt cutters from the backpack.

My legs moved before my brain caught up. I pushed past a waiter, who juggled plates with a hissed curse I didn't pause to acknowledge.

By the time I reached the door, Paul, ever the artist, but never fragile, had a grip on one thief, fingers clenched in the fabric of his hoodie. The other figure, the one with the backpack, slid out a knife, his gaze fixed on Paul's back.

'Oi, muggins!' I shouted, my voice cutting through the air like a whip.

The figure's head whipped around with a flicker of hesitation, that was all I needed. I squared my back foot and planted my weight, the rubber sole of my flat gripping the pavement, and swung a kick sharp and hard into his calf. He crumpled with a gasp, the knife clattering away and skittering under a parked car to our safety.

I shifted into a Muay Thai stance, knees loose, ready for whatever came next. Liam's patient, insistent training flooded my limbs, a muscle memory of feints and footwork, balance, and readiness. The figure staggered upright, clutching his leg, his eyes wide with pain and disbelief. I moved like Liam taught me, a quiet, coiled menace, and looked sharply into his eyes. Whatever fight he had left evaporated. He turned and stumbled away, weaving through traffic, the hindrance of fear outweighing greed or retaliation.

In the commotion, Paul's grip must have faltered; his perp twisted free and bolted down the street. Adrenaline still wandered in my veins, but the moment, thankfully, was over. The street began settling back into its basic rhythms, to a usual tide of people, mostly oblivious or uncaring.

Paul turned back to me, breathless and still rattled, a spark of awe in his eyes. 'Hell of a takedown... Ninja D.'

I smiled, feeling the tension dissolve. 'Liam taught me to keep moves real, and drilled me relentlessly while wearing padded gear. Comes in handy in a world that gets a bit... unpredictable.'

He laughed, shaking his head. 'That dance of yours, no-nonsense and sharp as hell.'

A warmth spread through me, the echo of Liam's voice still clear. "The spook-em" Thai dance, he called it. Perfect for scaring off amateurs. I glanced down, thankful I wore my flats today, which had kept me grounded and dignified rather than ending up on my backside in heels.

We then made our way back into the cafe, to a few approving nods and a 'Good on ya, lass..' following us. We reclaimed our seats, the table still holding the warmth of our earlier conversation. Paul's excitement bubbled up again, undimmed. We didn't let a couple of desperate thieves ruin the day or the meal. Still, a shadow lingered, a quiet sadness for how brazen, how desperate, these acts had continued to evolve.

When we finally parted, Paul set off with a new buoyancy, the wind in his favour, the road ahead clear. I turned back toward work, the weight of reality pressing in again, though something in me felt more focused, more sure. London shapes you that way. The constant swing and pull between chaos and stillness wears on you, but it also sharpens you. Keeps you grounded. keeps you awake.

The museum received me like an old friend, its stillness a cool hand against the heat of the day. After that wild, breathless current of time spent with Paul, this was a place where edges softened, where light fell in measured pools, soothing and deliberate. I wandered through the quiet corridors, past glass cases where history lay captive. My footsteps trailed behind me, like faint ghosts in the still air..

In the vastness of the exhibition hall, Emma leaned over a display plan, the pencil behind her ear forgotten, her focus absolute. As

I neared, she lifted her head, the light catching the glass of her spectacles, a flicker of curiosity sharpening her gaze.

'You're back,' she said, pushing a stray curl out of her eyes. 'How was lunch with Paul? You look like you've just had a workout.'

I leaned against the table, letting out a breath. 'Lunch was... eventful. You wouldn't believe it. Good news: Paul's artwork is getting a major exhibition. Breaking news: we caught a couple of thieves trying to nick his bike. And had a bit of a melee outside the caff.'

Her eyes widened. 'You're joking!'

'Wish I was.' I exhaled. 'Let's just say Liam's training came in handy. Both legged it. Well, one of them, half, legging it with a badly bruised calf.'

'Remind me never to mess with you,' she said, shaking her head in amused disbelief... 'London's finest vigilante curator!'

Before I could respond, my phone buzzed in my pocket. I fished it out, the screen lighting up with Molly's name. A warmth filled me at the sight of it. Molly, the heart of the neighbourhood, always with a kind word and wisdom to spare. But that warmth flickered and died the moment I heard her voice.

'Hiya, my D,' she said, the usual lilt weighted down, brittle at the edges. 'I'm feeling banjaxed. Éireann's in hospital, on the psych ward. She... she tried hurting herself.'

My stomach dropped, the world tightening into a sharp point. 'Oh, Molly...Where is she?'

'She's in the Royal. Been there since last night. She just rang, sounded drowsy but stable enough.' There was a pause, a tremor in her voice. 'I'd shut up shop in a heartbeat and go see her now, but I've got eight orders to go out and no one here but me.'

'Don't worry, Molly, I'll get there as soon as I can.'

Her breath hitched with relief. 'You're an angel, my D... If you can get in, I'll follow as soon as I shut up shop, like.'

'I'm on it, Molly,' I said, already reaching for my bag. 'I'll be there soon.'

'Thanks, love. You're a godsend.'

The call ended, leaving a hollow silence in its wake. I took a breath, steadying myself, then turned to Emma.

Emma must have seen something in my face because she was by my side in seconds. 'What's wrong?'

'I need to go," I said, shoving my phone into my pocket. 'It's Éireann. She's in hospital. Tried to hurt herself.'

'Don't even think about it,' she said, waving me off. 'I've got this covered. Go.'

I squeezed her shoulder in thanks, the pressure of her understanding grounding me. Then I slipped out of the museum, the calm, curated world behind me fading as reality surged back in, urgent and raw.

The streets seemed sharper and more defined, the weight of the situation pressing at the edges of everything. I walked quickly, weaving through the bustle, my mind already with Éireann, fragile, vulnerable Éireann. I thought of Molly, holding everything together with grit and frayed determination.

The city pulsed on around me, uncaring and relentless. But I pushed through. Some battles you choose. Others you answer without question. This was the latter.

I found my way through the maze of hospital corridors, the smell of antiseptic rising around me as I weaved through wet-floor signs. The hum of a floor buffer muffled conversations, as my focus narrowed as I approached the ward. A nurse behind the desk glanced up, gave me a nod and gestured toward a door down the hallway. My steps quickened firmly, heading towards my destination.

When I reached the room, I paused for a heartbeat, my hand on the cold metal handle. Bracing myself, I pushed the door open.

Éireann was there, curled up on the narrow bed, her arms wrapped around her knees tightly. The pale light filtering through the half-closed blinds cast shadows on her face, making her look smaller, more fragile than I'd ever seen her. Her eyes met mine, and the tears flowed immediately, raw, heaving sobs that shook her frame.

I stepped in without hesitation, closing the door behind me. In a few strides, I was at her side. I sank onto the bed, wrapping my arms around her, and pulling her close. Her body trembled against mine, and I rocked her gently, my chin resting on the top of her head.

'It's okay,' I whispered, my voice steady and low. 'It's okay. It's okay, sweetie.'

I held her tightly, feeling her sobs ripple through both of us. I had to force my own tears back, swallowing the ache that threatened to break free. She needed strength right now, not more sorrow. My fingers stroked her back, the gesture more for me than her, to remind myself to stay grounded.

After a while, the sobs quietened to shuddering breaths. I loosened my hold slightly and leaned back, meeting her swollen eyes.

'Okay,' I said, keeping my voice calm and measured. 'Tell me what happened..?'

She sniffed, wiping her nose with the back of her hand. 'I got asked to leave the flat. Well behind in the rent I was.' Her voice cracked. 'I worked so hard, D, but I couldn't keep up. Everything just got too hard. My head went.'

I closed my eyes for a moment, the weight of her words settling on my shoulders. 'I'm not surprised, in some ways,' I said gently. 'Last time I saw you at Molly's shop, you looked done. It had me concerned. I told you to call if you were battling.'

A nurse wandered in, her eyes flicking over the machines and notes. She gave us a polite smile, jotted something down on her clipboard, and left as quietly as she'd come.

Éireann looked away, shame tightening her features. 'I know, D,' she murmured. 'But I didn't want to burden you.'

I let out a slow breath, waiting for the door to click shut behind the nurse. 'Okay. So, Molly said you tried hurting yourself?'

She nodded; her gaze fixed toward the floor. 'I drank myself into a stupor and... I don't want to talk about it. But my flatmate called an ambulance. And now I'm here.'

Her voice broke, and she began sobbing again, her words tumbling out in gasps. 'I just didn't want to be here anymore, D, I didn't want to be no more.'

I squeezed her hand, the fight to hold back my own tears, a war inside me. I tilted her chin up so she could see the resolve in my eyes, the unshakeable support I needed her to feel.

'Now stop that thinking,' I said firmly, my voice soft but unyielding. 'Molly and I love you too much. You're not going anywhere.' I gave her a smile, strong, steady, a beacon I hoped she could hold onto.

She held my gaze, the storm behind her eyes flickering, dimming just a little. And I kept holding on, because sometimes, holding on was all we could do.

I waited until Éireann's breathing softened into a fragile rhythm, the weight of her exhaustion pressing into my arms. Stroking her hair gently, I whispered, 'How long do the doctors want you to stay here?'

She sniffled, her eyes glassy and red. 'Two more nights,' she murmured, her voice thin and fraying, barely holding itself together. Her lips trembled under a fresh wave of tears that welled up. 'But after that... I've got nowhere to go, D.'

Her words pierced through me, each syllable sharp and painful. I tightened my hold on her, a surge of fierce protectiveness rising, like a mother bear over her cub.

'Yeah, well, they're understaffed, overloaded, and not part of the feckin military budget,' I said, my voice low but firm. 'But it's okay, sweetie. I started mapping out a recovery plan for you on my way here.'

I leaned back just enough to look into her eyes, making sure she saw the certainty in mine. A strong, unwavering gaze, the kind that doesn't falter. The kind that promised she wasn't alone.

'I'll be back tomorrow,' I promised, my words calm and resolute. 'Then you're coming home with me the day after.'

She didn't speak, but she didn't need to. Her silence curled around me like a fragile bond of trust, her eyes glistening watery with a faith that she didn't have to voice aloud. It was as if I was carrying her across some rickety suspension bridge suspended over a chasm of despair, a bridge that creaked and swayed beneath us, but she held on, believing I'd guide her to safe, solid ground..

For a while, we spoke in halting tones. Eventually, our conversation drifted to the complicated knot of reasons why she couldn't turn to her family. As she spoke, her voice tripped and stumbled over memories that refused to smooth out, jagged and raw. I didn't try to mend them. I simply held space for her. When her words finally fell silent, I whispered, 'It's going to be okay. We'll figure it out.'

I stayed until the tension in her shoulders eased, until the haunted edges of her eyes softened lightly. When I felt she'd gathered enough strength to face the night, I kissed her forehead and slipped away.

The walk back to the flat felt longer than it should have. The city's night air was crisp, but the weight in my chest made it feel leaden. When I finally opened the door, the fond familiarity of the flat wrapped around me.

Brit lay sprawled on the couch reading, a lazy grin on her face, appearing from behind its cover. 'Look who's home,' she said, her eyes glinting. 'Dancing dragon!'

She stretched, a casual shift of limbs, then sat up. 'Paul was rambling about the Great Battle of the Bike,' she said with a smirk. 'But you know Paul, his phone died halfway through the story.'

I smiled faintly at the familiar chaos of it all, but it didn't reach my eyes. Brit's gaze sharpened, her teasing demeanour fading as she noticed the solemnity etched into my face.

Her brow knitted into small creases. 'What's wrong?'

I let out a slow breath, my bag slipping from my shoulder to the floor. 'It's Éireann,' I said, the name alone carrying the weight of everything left unsaid. I filled Brit in on the details over a good cup of tea, ending with, 'She needs a place to stay. Can she crash here for a while?'

Brit didn't hesitate. Her nod was immediate, solid. 'Of course. We'll figure it out. Whatever she needs, D.'

Relief seeped in, cracking the heavy armour around my chest, knowing Brit would be a fortress. 'Thank you, once she's settled, I'll take her back to Ireland. We'll figure things out from there.'

Brit reached out, swaying me gently in a reassuring hold of friendship. 'We've got this, D. She's going to be okay.'

I held onto those words, and Brit, letting them anchor me. Maybe sometimes, that's all it took. In the end, perhaps that's all we ever needed, to hold fast, to believe, to carry each other, as women bound by something deeper than words do.

Chapter 8
The Threshold of Forgotten Prophecy

The day had been long and arduous , exacting its toll. Exhaustion pressed heavily on my eyelids. I decided to turn in early, but first, I scrolled through the forum to wind down.

Anna had posted a video: Indigenous Native American Prophecy (Elders Speak, Part 1), narrated by Floyd Red Crow Westerman. The steady cadence of his voice and the solemnity of his words grounded me. A particular line echoed somewhere deep within me, striking a chord I didn't realise was taut.

"*Everything is coming to a time where prophecy and man's inability to live on earth in a spiritual way will come to a crossroad of great problems. It's the Hopi belief, it's our belief, that if you aren't spiritually connected to the earth, and understand the spiritual reality of how to live on earth, it's likely you will not make it.*"

Even though I'd always taken prophecy with a healthy dose of scepticism, this stuck. The words settled over me like a fine mist, silent, inescapable. Maybe it was the weight of the day pressing down, or a crumbling world outside, the relentless news cycle that made those words hit differently. Something aligned within my being. A whisper, delicate yet insistent, stirring in the depths, resonating louder than I was willing to admit.

Feeling deeply weary, I closed the forum and crawled into bed. As sleep tiptoed in, I found myself drifting to thoughts of taking Éireann back to the cottage. The idea of that quiet sanctuary, far away from the chaos, felt like a lifeline.

I was halfway to sleep when a fragment of a dream from Mum rose to the surface - the vision of a great tree. Then, out of the half-light of my mind, I heard something almost audible: "*One Male.*" The engraving on the hilt-guard of the Celtic sword that Emma and I had puzzled over flashed in my mind. My eyes snapped open, a thread of connection forming.

Nostradamus's *Vaticinia Code*, otherwise known as the Lost Book of Nostradamus.

I'd seen a vague image of it in a documentary ages ago, but that page, the plate depicting the "*Book of Life*", with a tree on one side and "*One Male*" written starkly on the other, had lodged itself in my subconscious. It surfaced now, clear as day.

Reaching for my phone, I realised I hadn't updated Emma on Éireann. Amidst the hectic fog of the day, I'd forgotten. Emma was my go-to for things like this, a walking archive of knowledge and curiosity. If anyone had insight on the Lost Book of Nostradamus, it was her.

I tapped out a message, briefly telling Emma that Brit and I had Éireann under our wings, that we were making plans to help her through this mess. Emma's response was immediate and warm, her empathy radiating through the screen. Then we switched gears to the Vaticinia Code.

Of course, Emma had already delved into its origins. She then explained how scholars were still divided on whether the manuscript truly came from Nostradamus himself. But Emma, ever the sleuth, leaned towards authenticity.

She told me how the manuscript had been discovered in 1994 by Enza Massa, who was researching in the Italian National Library in Rome. The manuscript was dated 1629, and inside, the name "*Michel de Nostredame*"—Nostradamus's real name- was scrawled in a way that hinted at legitimacy. Though 1629 was long after Nostradamus's death, Emma pointed out a key detail: Nostradamus's son, César de Nostredame, had died earlier that same year.

Her theory was compelling. César, perhaps aware of his failing health, had chosen 1629 as the moment to present one of his father's most enigmatic works to Rome. The timing seemed deliberate, a way to safeguard the prophecies within the Vatican's archives. Nostradamus himself had been terrified of persecution by the Inquisition, so holding the manuscript back until the right moment made sense. Emma's explanation painted a vivid picture, a son's final act to secure a place for his father's prophecies within the Vatican's sacred library.

I thanked her for the insight, feeling something settle, like a door clicking shut, leaving me intrigued and a little awed. The call ended, sleep came slowly, my thoughts still tracing the delicate threads—Éireann's cautious trust, Brit's steady, unshakable presence, Emma's quiet, ceaseless knowledge. Outside, the world might have been teetering, bending under the strain of its own making, but for now, just for this moment, everything was still. And as my eyes closed, all things, worries, wonders, the quiet hum of uncertainty, were gently laid to rest.

I awoke the next morning, surprised by a sense of strength and vigour that pulsed through me. Perhaps it was the stirrings of a deeper instinct, a call to steady myself for Éireann, to be a torch-bearer of solid ground she needed right now. Or perhaps it was the strange realisation that certain dreams, coincidences, and fragments of understanding were slotting together, forming a mysterious but discernible puzzle. A puzzle slowly revealing itself, like the budding leaves of spring.

The maze on Zeus's arm from my dream no longer felt like some abstract motif, lost to dream logic. Its shifting pathways seemed a reflection of reality's own fluidity, a mirror of unfolding events. For the first time, it felt comprehensible. Even if my mind shook it off as a mere coincidence, the labyrinth was making sense, yet mysteriously.

Outside, the world was a wild symphony of rain. Heavy, furious drops of rain ricocheted off Brit's deck in a dance of chaotic rhythm. I stared at the downpour, the kind of rain that insisted on itself, and decided not to let it dictate the day's energy. Instead, I set myself on the exercise bike, determined to embrace the storm's ferocity with motion and sweat. Rainy days like this called for rhythm. For something to propel me forward.

I queued up a forum-recommended track that struck a deep chord within me—*The World Turned Upside Down* by Billy Bragg. A song that was itself a history, a protest woven into melody. A ballad of the Diggers, those 17th-century visionaries who dared to cultivate common land, led by the resolute Gerard Winstanley. A ballad of the Diggers, those 17th-century visionaries who dared to cultivate common land, led by the steadfast Gerard Winstanley. They envisioned a world of shared prosperity, where wastelands bloomed into sustenance. But their dream was crushed by the authorities who drove them from St George's Hill, a hill now, with bitter irony, transformed into a golf course surrounded by affluence.

The music surged through my headphones, the beat seeping into my bones, everything syncing with the motion of the pedals, each word igniting something deep and persistent within me. A particular aspect of a verse asked me to rise from my saddle, like a battle cry.

I pressed on, pedals turned beneath me, the rhythm of my breath steady, almost like a pulse of its own. My breath then came in sharp, quick bursts, each push of the pedals pulling me deeper into the moment. Sweat beaded on my skin, its coolness marking the passage of time, as I focused solely on the motion. When I finally stopped, my body hummed with the quiet satisfaction of exertion, a satisfaction that carried through the stillness of the room long after the ride ended.

A cool shower washed away the sweat, but not the resolve. Coffee, dark and strong, grounded me further. I grabbed my broad, trusty umbrella and stepped into the pounding rain. The rhythmic drumming of the raindrops against the fabric accompanied me as I made my way to the museum.

By midday, the rain relented, but my curiosity over Nostradamus's Vaticinia Code only intensified. Over my lunch break, Emma and I turned our attention to the Vaticinia Code with renewed focus, beginning with Plate 67, the Book of Life. It felt like the epicentre, the linchpin of everything we were trying to unravel.

The image held my gaze longer than I expected. A tree, a hand, strong, poised, lion-like fingers curled in a way that hinted at both command and grace, a cryptic inscription on the page: "*One Male.*" Each element felt deliberate, yet subtle with meaning. The hand with leonine strength seemed to suggest. To truly understand the Book of Life requires knowledge, a natural ascendancy, and a strength of character. We surmised that the crown above the figure wasn't the mark of traditional royalty. Instead, it mirrored the crown of the Tree of Life itself, a symbol of interconnected wisdom rather than earthly power.

Most of the plates left us perplexed, each symbol a riddle, each detail a potential clue. Decoding them felt like chasing shadows, but there was something tantalisingly relevant about the imagery. A sense that history and evolving time were speaking softly, if we only knew how to listen.

Emma, ever diligent, traced Nostradamus's life, his fears, his secrets, the labyrinthine journey of the manuscript. We both agreed that interpretations of such works were speculative at best, and to avoid descending into the realm of wild conjecture, we tabled the discussion for another day.

As the afternoon waned, I called Molly to check on Éireann. Her voice, warm and familiar, carried a note of quiet strength.

'I took her some groceries and odds and ends,' Molly said. 'She's doing well, all things considered. Bless her. Told me you've taken her under your wing. You're a grand woman, D. It's steadied her no end from the day before.'

Relief of Éireann settling down eased into me, like a wave hugging the shore. 'All will be okay, Molly. Brit and I have it sorted.'

As I juggled the phone and tidied the day's scattered remnants on my desk, a shared sense of hope flowed between us. In hindsight, we both knew the precarious edge where Éireann had been standing, pushed, pulled, worn beyond the bone. But now, there was something like a path forward beyond despair, slender and fragile as new growth, yet real.

The call ended in camaraderie and eased minds. I gathered my things. Slinging my bag over my shoulder, I made my way down the hallway that stretched ahead, the soft echo of my footsteps a quiet reminder that, amidst the chaos, things can fall into place.

The next day, I went to collect Éireann. The rain had worn itself down to a sullen drizzle, clinging on out of habit more than intent. I moved through the corridors, past clusters of murmuring voices, the shuffle of rubber soles against linoleum, the mechanical pulse of unseen machines ticking out time. The lights overhead hummed, washing everything in a colourless, unyielding glow.

Then, stepping into Éireann's room, I stopped short. My breath caught, my hand raising to my mouth to stifle a laugh. The place had been transformed, no longer a sterile, impersonal space but something closer to a scene of controlled chaos. Groceries were everywhere. It looked as though a doomsday prepper's basement had detonated in the middle of the sterile hospital room. I shook my head, grinning despite myself. I was expecting an overnight bag, not a feckin farmer's market.... Tins of soup were stacked like miniature towers. A burst bag of oranges scattered its bright, rolling bodies across the floor like marbles.

Éireann's eyes sparkled with dry amusement as she shrugged. 'I know. Molly overdid it a wee bit. She swaggered in with bundles, and a burly friend in tow with even more..!'

'Bless her,' I replied, still blinking as I tried to mentally calculate the sheer volume of it all. My brain whirred with the logistics of getting it back to Brit's flat without some form of a hernia. 'How in the name of Jaysus are we going to carry all this?' I muttered to myself.

Éireann smiled with a quiet determination. 'Well, D, I'm used to a bit of graft. We'll give it a crack, eh..?'

'OK, let's get to it, Éireann,' I replied with a sigh, already wincing at the thought of the journey ahead.

We set to work, dividing the supplies between us. We stuffed bags, looped handles around wrists, while balanced awkward bundles against hips. By the time we were finished, we resembled a pair of pack mules, stooped and laden with provisions. We made quite the sight as we shuffled puffing through the hospital corridors, past curious onlookers and bemused nurses.

Somehow, we emerged out into the damp dusk air, the drizzle now feeling like a gentle mockery of our struggle. I glanced at Éireann; she was grimacing, but her eyes were alive with purpose. For the first time in days, her spirit seemed to burn a little softer and brighter.

We trudged onwards, making slow but determined progress. By the time we reached Brit's doorstep, my fingers were numb, and my shoulders were drained. I barely had the energy to lift my hand and knock.

Brit opened the door, her eyes widening as she took in the spectacle of us. 'Fek. Check this out, Paul..! The Irish sherpers have arrived.'

Paul appeared behind her, his eyebrows shooting up as he took in the mountain of supplies hanging off us. 'Jaysus, you two look like you've just come back from a right pillaging of the Silk Road.'

Éireann sighed, setting her load down with a theatrical groan. 'Close enough. This is Molly's idea of a few bits and bobs.'

I dropped my bags beside hers and flexed my fingers, half expecting them to stay curled permanently. 'I think my arms are about two inches longer now.'

Brit stepped aside, shaking her head with a grin. 'Get in before you keel over.'

As she eyed the haul, her mouth twitched with amusement. "I'm not sure if you two need a medal or a chiropractor.'

'Both, probably,' I muttered, rubbing my lower back with one hand and shoulder with the other.

As Brit and Paul helped us sort the supplies, I caught Éireann's eye. She gave me a quiet nod, the kind that said more than words ever could.

We'd given it a crack, all right. And we'd made it home.

Night settled in, slow and unhurried, wrapping Brit's flat in a kind of drowsy glow. Outside, the rain had lost its resolve, thinning to nothing. Brit and I set about making a proper dent in the haul of groceries, commandeering the small but efficient kitchen like a pair of well-practised generals, the kitchen, small but battle-tested, became our domain, steam curling lazily from bubbling pots, oil hissing as it met the pan, the quiet clatter of knives against cutting boards filling the air.. Our purpose was clear: a meal hearty enough to kick Éireann's recovery into gear.

I served Éireann an extra portion, nothing over the top, but enough to help boost her energy. She needed it, that was certain. Halfway through my own meal, I glanced up and asked, 'How is it ?'

But her answer was already evident: her plate was wiped clean, and her grin broad and triumphant as she exclaimed, 'Smashed it!'

I couldn't help but smile. 'Well, you'd better save room. Paul's on dessert duty tonight. Brace yourself for strawberries and ice cream luxury treatment around here.'

After dinner, she rose with unexpected determination, insisting she handle the dishes herself. Brit began to protest, 'No, no-oh, all right, if you insist' quickly relenting with a wry smile.' On the other hand, Paul stretched out dramatically and declared, 'Splendid! I'll take the night off, then.'

By the time Éireann had finished her self-appointed task, Brit and Paul had tucked themselves onto the couch, tangled together in that easy way only they could manage. The light from the TV flickered against their faces as they settled into some documentary. Éireann, her tidying mission complete, followed me to my room.

Once there, I sat her down, my voice soft but steady. 'Éireann, I need to know what they've given you. Meds, I mean..?'

Her response came slowly, as though testing the weight of her predicament. 'Antidepressants and some benzos for anxiety.'

I nodded, letting her words settle before speaking. 'All right. Listen, trust me on this. When I first moved to London, away from home, from family. I had a torrid time back when. A Doc prescribed me meds, honestly, it was the worst mental fog I've ever been in. Coming off them? Hellish.'

She watched me, her expression intent.

'Here's what I want you to do, ok,' I continued... 'Don't throw them away. Keep them, eh, but put them aside for a couple of weeks, just for now. As for the benzos, they're okay for the really rough moments, but only short term. Trust me, that's a pit you don't want to tumble into either. So many doctors hand this stuff out like sweeties, but it's like throwing a blanket over a car wreck and hoping it'll run again. There are no SHAZAM fixes in this world, Éireann. Just legalised drug dealers bathing in swimming pools of dosh, at others' expense.'

Her gaze softened, and for a moment, the room felt impossibly still. 'Okay, D. I trust you, no end.'

'Good,' I said, brushing a stray lock of hair from her face, letting her see the quiet certainty in my expression. 'You're safe here. We'll shelter you from the storm, all right?'

A small, grateful smile flickered across her face.

'Now, come on," I said... 'Let's go make a proper pot of British cha and join the couch doves.'

The night wound down with Éireann tucked up on a cosy fold out stretcher bed, safe and sound under the muffled rhythm of traffic curling over Brit's deck. I headed off to bed myself, sinking under the covers with a sigh of relief.

As I lay there, a strange sensation washed over me. It was like being wrapped in something both vast and tender, a veil of calm sending gentle tingles down to my fingertips and toes. It wasn't unpleasant, far from it. But it was curious, all the same.

An urge rose in me, sudden and insistent: check the forum. Sleep tugged at me, but curiosity proved stronger. I reached for my laptop, scrolling through until I found the Zeus dream thread.

A new comment caught my eye, posted barely a minute ago:

Intriguing vision, Ms. D. I may be able to offer some valuable insight, happy to go over it if you're interested. Feel free to DM me.

Best,

TR

The words settled into me like stones dropped into still water, ripples of déjà vu spreading outward.

Intrigued, I sent off a reply:

Hi there, TR. YES, I would be very interested in your thoughts about the dream.

Regards,

D

I waited, the laptop's soft glow bathing the room. Moments later, a reply appeared:

I'll get back to you. I just need some time to reflect on it further, and when the moment feels right, I'll reach out again.

Kind regards,

TR

I smiled faintly, replying with a simple:

That's fine. Looking forward to it, TR.

Kind regards, D

Anna hadn't responded yet, and sleep was pulling harder now. Closing the laptop, I slid it onto the side table and let my eyes drift shut.

That strange sensation returned, the tingling carrying me off like a boat made of feathers, gliding me gently into the night.

The clock read 1 am when it began. I wasn't merely dreaming. I was somewhere else entirely. Where before me stood a gateway, unlike anything I'd ever imagined. It wasn't just a structure; it was alive, fluid and celestial. Its surface shimmered luminous blue, reminiscent of Lapis Lazuli. The material was otherworldly. Drawn forward. I stepped through.

On the other side, I saw a man seated, hunched over, his head buried in his hands. Without uttering a word, I asked him. Have you been in agony?

His answer came not in words but in a vision, of a wound on his side, deep and ancient. It was more than an injury. It was a story, a mark left by something older than words. As I focused on it, time loosened, unspooled like thread slipping through fingers. The wound carried histories within it, shifting, overlapping, slipping between one moment and the next. A past within a past. Pain folding into sacrifice, sacrifice into something deeper. He bore it all, as though his skin were parchment and the myths of old had been written into him, layer upon layer, until his very being had become a living archive.

I stepped closer, resting a hand on his shoulder. 'Did you know that within you is the knowledge of all?' The words left me before I could think of them, slipping from somewhere beyond the edges of my mind.

And then everything shifted, and I found myself standing in an open field. The air was thick with the scent of earth, of something fresh and stirring. The wind moved lazily over the tall grass, bending it in slow rolling waves. And there, standing in the soft, dappled light, was Anna. She was just as I had seen her before, wrapped in the pale folds of a buckskin ghost dance dress, her dark eyes steady, filled with something deep and knowing. She smiled, not as a greeting, but as if she had been waiting. Then, wordlessly, she turned her head, her gaze pulling mine toward the hill behind her.

And there it was. Cresting the rise, its thick white fur shifting in the breeze, moving with an unhurried, deliberate grace—a great bison. It stepped forward, powerful yet weightless, as if the very earth recognised its presence. Each motion felt sacred, every shift of muscle, every breath.

Anna turned back to me. No urgency, just quiet certainty, and though no words were spoken, her message was unmistakable: It's time.

I reached for her. My fingers found hers, warm, solid, steady. And in that instant, the dream collapsed. I gasped awake. My chest felt heavy, as though I had carried the weight of the dream back with me. The weight of it lingered, humming in my veins, waiting to be understood.

Instinctively, I reached for my laptop and opened it, half-certain there would be a message from Anna, and there it was, waiting for me:

Thunder rolls in the distance, whispering what is to come before it arrives. D, it is as it should be. TR "Thundercloud" walks his path in his own time; he will reach for you when the moment is right, perhaps in a week or so. The river does not rush to meet the sea, yet it always finds its way.

Blessings,

Anna.

My hands trembled slightly as I typed my reply:

Thank you, Anna, but how did he know I'd ask you? And why *Thundercloud..?*

Her response came swiftly, as though she'd been expecting my questions:

He sees into people, and D feels the shape of their thoughts before they speak them. So, you reaching out? That was already written in the current. '*Thundercloud*', that name came to me in a vision, strong as the storm it carries. I'll share it with you soon, when the time is right.

Blessings,

Anna.

I replied:

OK, Anna. Thank you. Love and peace to you, too.

D

I closed the laptop and leaned back, exhaling deeply. The dreams, messages, and mysteries swirled in my mind. The weight of the dreams, the messages, the strange pull of it still clung to me, refusing to settle. But exhaustion was patient, insistent. I pulled the covers up, the fabric cool against my skin. The memory of that night lingered, sharp in places, blurred in others, like the last notes of outside traffic humming in the air. Sleep crept in, quiet and inevitable. Somewhere at the edge of it, a whisper remained.

Chapter 9
Whispers Before the Turning

Just over a week had passed since Éireann had landed at Brit's flat, and I couldn't help but feel a quiet relief; she'd slotted in with remarkable ease. She wasn't just recovering; she was becoming more present in the world, and more grounded, refreshed and carrying a spark of vitality that had been dulled for too long. Brit, ever observant, had already caught on. She noticed how Éireann had taken to the exercise bike with a kind of cautious determination. Being Brit, she seized the opportunity. 'Fancy a few days' work?' she'd asked, promptly roping Éireann into helping her wrap up a sprawling landscaping job she'd been grappling with.

On their final day of landscaping, Britain seemed intent on embracing climate change, with winter like a ghost, returning to haunt London in earnest. The air was biting, dampness clung like a second skin, and the city appeared bent on reminding everyone of its mercurial moods. Brit and Éireann returned to the flat late, looking like a pair of crab fishermen who'd lost their way back from port. Their wellies squelched on the floor, their wet gear hung off them in heavy folds, and their beanies, sodden and misshapen, did little to hide their weary expressions. They'd barely made it two steps inside when Paul, stretched out on the couch in his usual languor, gave them a once-over and smirked.

'Look, D,' he said, gesturing lazily. 'The crew from Deadliest Catch just walked in..!'

Despite their bedraggled state, Brit's mouth twitched upwards. 'Well, Moby best get off the Great faffing Couch,' she shot back, peeling off her coat with dramatic flair, 'Captain Ahab here and her deckhand need steaming cups of Java. So, go man the coffee machine, eh..!'

'Aye aye, Captain,' Paul drawled, pulling himself upright with an exaggerated groan. He shuffled off to the kitchen, but not without a wink that betrayed his amusement.

I grinned at the sight of them. 'Well, I assume you decided to finish off the task at hand, eh, me hearties?' I teased in a mock pirate tone, leaning against the counter.

'OK, OK,' Brit replied, rolling her eyes as she wrestled her boots off. 'Time to pull up the anchor, D. Next thing you know, you girls will be singing Irish sea shanties.'

Éireann let out a quiet laugh as she fought her own battle with soaked layers, her movements slow and deliberate, her exhaustion palpable. 'Aye, D,' she said, finally sinking into a kitchen chair with a groan... 'We got the job done, but I'll tell ya, Brit has no off button. There isn't a muscle in my body that isn't aching for a warm bath, gawd.'

Paul reappeared with two mugs of coffee, sliding one to Brit and the other to Éireann. I handed her a second cup for good measure, a double shot, because if anyone deserved it, it was her. She took a long sip, cradling the warmth between her hands as though it were a lifeline.

'Well,' I said, giving her a knowing look, 'when you were feeling good about doing a couple of days' work with Brit, I did warn you, you were in the eye of the storm. Brace yourself..!'

Éireann winced, the warmth of the coffee softening her edges. 'I hear ya, D,' she murmured.

I waited until she took another sip before leaning casually against the table. 'Anyway, rest up tonight and tomorrow,' I said. 'I've got news, eh! I've sorted out some time off from the museum, and we're heading to Ireland.'

She froze mid-sip, her eyes darting to mine. 'Ireland..?' she repeated, her voice low, as though testing whether she'd heard correctly.

'I've spoken to my sis, Sinéad. She needs a bit of help at the pub. I thought you might like to pitch in for a while. Now, how's that sound to you?'

Her face changed in an instant as the fatigue seemed to melt away, replaced by something brighter, hope, maybe, or the first glimmer of something grounded. 'Grand,' she whispered, her voice catching. 'Oh, grand, D.' But her words wavered, and her eyes glistened with unshed tears.

Before she could falter, I stepped forward and wrapped her in a hug. 'You don't have to say anything,' I said, my voice steady. 'We're not finished with you yet. Solidly mended, that's the goal, eh?'

She nodded into my shoulder, her breath hitching slightly. Before the moment could get too heavy, Brit bounded over, throwing her arms around the two of us with an exuberance that defied her exhaustion. Straightening her back and stiffening her torso, she lifted her chin and launched into a ridiculous imitation of Irish step dancing. Her arms dropped rigidly to her sides, and her legs kicked out in chaotic, mismatched moves. 'I'd come with you ladies,' she declared with mock solemnity, 'but Ireland's tis jealous of my cracking dance moves..!'

Éireann let out a laugh, her earlier emotion bubbling into something lighter, freer. I couldn't help but join in, the laughter pulling us forward, dissolving the heaviness of the past week. Brit, her antics unstoppable, nearly knocked over a chair with her exaggerated jigging, which only sent us into further fits of laughter.

It was a scene far removed from the sombre shadow that had hung over us when Éireann first arrived. Now, with laughter filling the flat and a new adventure on the horizon, it felt like we'd turned a corner. Better days were just ahead, close enough to touch.

Apart from Éireann's steady improvement, the week had carried a weight. I couldn't shake an undercurrent of thoughts that seemed to linger like an old scent, always just out of reach. With its strange clarity, my last dream had kept resurfacing, and then, there was that encounter with TR, cryptic as ever, brushing against the edges of my mind like a shadow passing through the dimmest light. Anna's words followed me too, curling around the stillness of my days, threading themselves into the quiet spaces between tasks and thoughts. That evening, as I often did, I logged onto the forum. It had become something of a ritual, a peculiar comfort, a place where the boundaries of the unexplainable were blurred, and the fragments of dreams, visions, and mysteries could be laid out, flickering like stars in a sky just beyond reach.

But this week, something was different. The dream-vision section was alive, more alive than I'd ever seen it. The posts were pouring in, vivid and raw. While none mirrored my own experience, they carried a shared essence, a feeling like threads drawn from the same unseen web. It wasn't obvious at first glance, but it was there, subtle yet undeniable.

I clicked through a few, my curiosity growing. Was this just a coincidence, or was something shifting? A movement, perhaps, like the first notes of a melody trying to find its shape. But before I could linger too long, I checked my messages, my heart fluttering gently.

I didn't know what I hoped to find, a reply from TR, perhaps. Something more than an acknowledgement, something deeper, resonating with the Zeus dream or this growing sense of interconnectedness. But there was nothing from him. Instead, I found a single new message from Anna- "White Hawk."

My heart leapt a little as I opened it:

Hi D, About TR-"*Thundercloud*"

The winds have not been kind to me, and I've been held back by the weight of family, the sickness of my mother, pulling my focus to the ground. But I'm here now, and the vision I carry is too strong to leave behind.

It came to me, the spirit of Tȟašúŋke Witkó—*Crazy Horse*. He stood before me, a presence immense, towering over all. His horse beside him was covered in symbols that twisted and shifted as if alive, moving with a strange playfulness. His eyes, his power, were beyond human, beyond the earth itself. And yet, he was there, with me, leading me.

And then, from the depths of the vision. I saw a boy. Small against the storm that was building on the horizon, but unmoving. Crazy Horse walked toward him, and with each step, the clouds began to churn, rolling dark, a storm vast enough to swallow the sky. The boy stood, quiet, as Crazy Horse opened his arms to the clouds, as if to embrace the fury itself.

Then the boy smiled. A quiet, knowing smile. And from him, from within him, something stirred, something great, something beyond. A Thunder Being, Wakíŋyaŋ, the Lakota call him. Not one man, not one people, but all nations. He rose from the boy, immense, a force of nature. His hand brushed the clouds, and the lightning played along his arm, wild and fierce. He and the boy were one, a connection not just of body, but of spirit.

The boy reached up, and as his hand touched the sky, the storm broke open. A rain began to fall, not a rain, a sacred healing rain. It shone, radiant, like the light itself had turned to water. And I knew, deep in my soul, this was not just any rain. It was for the Earth. For all the Earth. The storm, the lightning, the rain, everything was of one, and that one was the healing we have forgotten.

I wept as I felt a nearness of the Seventh Generation Prophecy. Crazy Horse's prophecy, the words echoed in me:

"*Upon suffering beyond suffering, the Red Nation shall rise again, and it shall be a blessing for a sick world... A world filled with broken promises, selfishness, and separations. A world longing for light again. I see a time of seven generations when all the colours of mankind will gather under the sacred Tree of Life, and the whole Earth will become one circle again. In that day, there will be those among the Lakota who will carry knowledge and understanding of unity among all living things and the young white ones will come to those of my people and ask for wisdom. I salute the light within your eyes where the whole universe dwells. For when you are at that centre within you and I am in that place within me, we shall be as one.*"

TR, whom I call "Thundercloud," invoked this vision. The boy, he was TR as a child. I'm certain of it. That's all I can say for now.

Blessings D, Love,

Anna

I stared at the message, feeling as though the air itself had shifted around me. Her words carried a weight that pressed gently but firmly on my chest, as though I were standing in the path of an ancient wind.

The vision was extraordinary. It spoke of connection, prophecy, and healing. And yet, it felt like only a fragment of something larger, as though I'd been given a glimpse through a narrow crack in the wall.

I steadied myself, absorbing it all, then wrote my reply:

Dear Anna,

Firstly, I hope your mother's health is improving. If there's anything you need, please let me know. As for the vision... What can I say? It moved me deeply and left me in awe. Thank you, Anna, for sharing something so profound and personal.

Love,

D

Anna wasn't online, and I didn't expect a reply anytime soon. Instead, I turned my focus back to the forum, a strange new energy stirring in me. I decided to write a post about Crazy Horse, delving into his prophecy and the meaning behind his name.

According to Black Elk, the name "*Crazy Horse*" came from a vision. In it, Crazy Horse saw that everything in the world was made of spirit, rocks, grass, trees, all of it. Nothing was hard or fixed; everything floated, everything alive and connected. His horse, too, moved strangely, not crazily but in a way that defied ordinary understanding. His name was less about madness and more a reflection of his horse behaving peculiarly.

As I wrote, I realised how closely Anna's vision mirrored this. It sent a shiver through me, like stepping into a river and feeling the undercurrent tug at your feet.

I posted my thoughts and switched off the screen, the weight of it all lingering in my chest. Sleep tugged at me, but it wasn't the kind that would settle easy. I had a trip to Ireland to prepare for, with Éireann, packing, planning, and sifting through the mess of it all. The forum, the visions, and even TR would have to wait. The light clicked off, but there was something still buzzing at the back of my mind. I knew this wasn't the end. It felt like a door half-open leading into a vastness I couldn't yet touch but could feel just beyond my grasp.

Chapter 10

Through the Veil and Home Again

The morning of departure to Ireland had arrived, and as I made my way into the lounge, the faint light of dawn spilling across the room welcomed me. I shuffled into the kitchen, bleary-eyed and in desperate need of a weighty cup of eyelid opener.

Éireann was already there, perched on the edge of the couch as if poised to charge through a boarding gate. Her bag sat neatly by her feet, and her expression was one of readiness and eagerness...

'What's the craic..? You're packed two hours early,' I mumbled, still half-drowned in sleep. My hand moved instinctively to my bed-head, scratching at the tangled remains of the night as I shuffled to the kitchen.

'I'm an early bird, D,' she replied, her voice light, as though this were any ordinary morning. But the way she smiled, wide and slightly uncertain, hinted at a buzz beneath the surface. 'Got everything sorted.'

I raised a sleepy eyebrow, offering her a half-smile as I turned towards the kettle. 'Pull up on the reins, lass; no need to rush the gate. Tea or coffee..?' I asked, my hand already measuring oats into a bowl.

'Tea, thanks,' she said, her shoulders softening.

The kettle hummed into action as Brit wandered in, her hair sticking up in all mad directions. She didn't say a word, simply drifted towards the deck with the tea I handed her, like some quiet ritual. I let her be. Brit awoke on her own terms, and I wasn't one to break that sacred silence.

Breakfast unfolded with a bit of banter, and before long, we bundled ourselves into Brit's little van. She drove us to Heathrow, navigating the traffic like the Londoner she was, swift and decisive.

At the airport, the goodbyes were quick but meaningful, a squeeze of Brit's hand, a smile exchanged in lieu of words. Éireann and I made our way through the labyrinth of check-ins, security, and gates until finally, we were seated on the plane.

The engines whistled to life, that familiar hum of potential energy building into the light, steady pull. I've always loved that bouncy building surge as the pilot hit the throttle. It feels like potential springing into life's adventure, a sense of forward momentum that evokes the traveller's soul.

As the plane lifted into the sky, Éireann reached over and took my hand.

'Home, here we come,' she murmured, her words soft but steady.

Her fingers were warm but carried a trembling undercurrent. It wasn't fear, it was something else, like a burden slipping away. All that had weighed her down and drained her for too long was loosening its grip.

'OK,' I said, breaking the silence to steer the mood forward. 'Sinéad will have you trained up on bartending and whatever else she ropes you into. You'll be in good hands.'

'And Maverick-van will be waiting for you,' I added, a grin tugging at my mouth.

Her brow lifted. 'Maverick-van?'

'It's a van Brit and I kitted out from scratch to go around Europe in,' I said, leaning back in my seat. 'In hindsight, we should've paid more for a lower mileage project, but after some learning curves, we turned it into a proper home away from home. Cosy, well-insulated, the works. Luckily, we made it back to Galway before the motor went kaput, not in the middle of Spain, thankfully.'

She laughed softly, and I continued, 'It's parked out back at Sinéad's pub now. Didn't have the heart to sell it. You'll have a roof over your head, rent-free. Sinéad even ran a power lead to it, though it's got solar.' I gave her a knowing look. 'Not that solar's much use in Irish weather.'

Éireann blinked rapidly, her voice catching as she said, 'D, I don't even know what to say. You've already done so much.'

I squeezed her hand, grounding her. 'Don't go crying now, Éireann, or the flight crew'll start fussing,' I teased, nodding towards the gap in the seat ahead where someone was peeking through, curious.

Her lips parted as though to speak, but I cut her off gently. 'Hush now,' I said, gesturing towards the window. 'Look, there's Ireland!'

Through the clouds, green fields emerged like an ancient tapestry unfurling below us, the edges glowing faintly against the dull grey of the sky. The sight was timeless, bound in its own magic.

Her breath hitched, and her fingers tightened on mine. 'Ain't it beautiful?' she whispered.

'Welcome home,' I said softly.

Éireann turned back to me, her face alight with something fragile yet certain. And in that moment, the word home seemed to mean more than just a place. It felt like a possibility, foundations yet to be built but steadily taking shape.

Not long after flying over Limerick, we touched down at Shannon Airport. The landing was smooth, almost anticlimactic after the buzz of the journey so far. Soon enough, Éireann and I found ourselves weaving through the terminal, the faint hum of conversation and the occasional bark of a tannoy filling the air.

The airport held its own rhythm, bright, unblinking lights, the low murmur of arrivals and departures, suitcases dragging, voices rising and falling like waves. Éireann moved ahead, already scanning the carousel, sharp and determined, with a focus set on spotting our bags. I let her. I lingered, letting the moment settle over me like the hush before a storm.

Then, like a tide swelling before it crashes, it hit. A force so vast and full, it nearly took my breath. Love. Not a thought, not a memory, but something bigger. A presence, a weightless flood moving through me, settling deep. It felt like recognition, like remembering something I hadn't known I'd forgotten.

I steadied myself, stealing a glance at Éireann as she fished for our bags, completely unaware of the ground shifting beneath my feet. Éireann tugged at a suitcase, oblivious. My hand found my pocket, moving before I had even thought to check, as if guided by something beyond habit. I knew before I saw it. The screen glowed. An email. TR had replied.

How did I know? It felt obvious, like knowing the sun would rise. But I didn't check. Not yet. Some things aren't meant to be picked apart under airport fluorescents. Some things need quiet. I'd wait. The cottage would be waiting, too.

Bags in hand, we weaved through the slow-moving crowd and found our way to the bus for the journey to Galway. We climbed aboard and sank into our cosy seats, the hum of the motor blending with the soft murmur of passengers around us. Outside the window, the rolling hills and patchwork fields of the Irish countryside unfurled like a painting, framed by dry stone walls and storybook towns that seemed to belong to another time.

What I enjoyed most, though, was Éireann. She leaned against the window, pointing out everything from the ancient stone walls tracing the fields to the occasional thatched-roof cottage nestled between the hills. There was something pure about her excitement, childlike, almost, as though she were rediscovering a part of herself long forgotten.

Fifteen minutes from the bus station, I called Mum to let her know we'd soon arrive. True to form, she was waiting as the bus pulled in, a quick wave from her catching my eye. Her china-blue eyes sparkled with warmth, and her smile followed, a kind of reassurance that home was exactly as I remembered.

As Éireann and I crossed the pavement to meet her, Mum's warm smile spread wide, the kind of genuine ease that made the world feel less jagged.

'There ye are, my two adventurers!' she called, her voice carrying over the light din of the station.

We bundled ourselves into the car, the boot packed efficiently with our bags, and Mum wasted no time in chatting. Her light banter filled the space, easing Éireann into a casual state of mind.

'So, you've survived travelling with this one, then?' she teased, gesturing to me with a grin.

'Barely,' she replied, catching Mum's playful, warm energy.

We left the station behind and made our way, as Mum continued on with her banter. It was her way of welcoming Éireann, making her feel at ease. It worked, too—Éireann's laugh came easier now, and the lines of worry on her face softened with every passing mile.

The roads grew narrower, the houses fewer and farther between, until finally, we turned onto the lane leading to my childhood home. The familiar sight of it came into view, nestled among the trees, its whitewashed walls glowing faintly in the fading light of the evening.

'This is it, and here we are, ' I said softly, more to Éireann than Mum, as the car rolled to a stop.

She leaned forward, her eyes wide as she took it in. 'It's lovely,' she sighed.

Mum turned to her with a grin. 'Well, let's get inside then. Come on, let's get you settled!'

We stepped out into the cool air and gathered our luggage. As we approached the front door, I felt that same strange wave of emotion again, a grounding warmth that welcomed me back to where memories began.

As the front door swung open, Puddles was the first to greet us, padding towards us. He looked up at me, his tail giving a calm swish, and let out a rattling chirp followed by a few melodious meows before heading straight for me. I crouched to scoop him up, the vibration of his purring, like the steady hum of a distant engine, spreading warmth into my very bones.

'Mum, you've gone and fed him into a weighty ball again,' I said, half laughing, though laced with a good hint of concern.

Mum gave a sheepish shrug, clearly unrepentant. 'I don't know what it is, my girl. He must just be more relaxed here, eh?'

I gave her a knowing sideways glance. 'Eh,' sure, Mum. He's been sleeping in the cat bowl again, hasn't he? Probably because you keep it filled to the brim. Look at him..!' I cradled him, the evidence undeniable in his contented girth.

'Well, all right then,' she said, waving me off with a hasty smile and a quick change of subject. 'Now, let's have a nice pot of tea.' Already she was retreating, the light of the kitchen drawing her forward. 'Your father's out in the sunroom waiting for you.'

'Okay,' I called after her, lowering Puddles to the floor. His tail swept lazily, a punctuation to his satisfaction. With Éireann beside me, and Puddles waddling closely behind, we made our way through the lounge.

The sunroom unfolded before us, its light soft and familiar, and there was Dad, nestled comfortably in his favourite old deep chair, as I'd expected. Though the signs of Parkinson's were present, I could tell it hadn't progressed much since my last visit; I could sense it in the way he greeted us, steady but still warm, his voice carrying that calm fortitude I'd always admired.

'Hi there, Clover. You're looking fighting fit, lass,' he said, his smile broad, his words warm. 'Reminds me, your pedalling machinery is in the spare room. The box arrived three days ago.'

I bent to embrace him, his arms strong despite the toil. 'Grand. How've you been, then..?'

'No complaints, getting along fine,' he replied, his ever-optimistic tone unwavering. Dad had always been one to find a silver lining, no matter how dark the storm.

I turned to introduce Éireann, her smile shy but kind, and soon we were gathered at the table. Two pots of tea stood sentinel among the plates of biscuits, their warmth rising in faint tendrils. The hours blurred as we shared laughter and stories, the room filling with a type of kindred spirit that stretches beyond words, spilling over and leaving no space for the struggles of life to intrude.

But time, persistent as ever, crept forward. I glanced at my watch, then at Éireann. 'We'd better get this one to Sinéad's place,' I said, rising reluctantly. "And Puddles and I need to get ourselves back to the cottage."

We packed the car efficiently, the exercise bike in its box went on the back seat, Puddles in his carry pack sat snugly beside it, and the luggage found its place in the boot. After hugs all around, Mum and Dad waved us off from the drive, their figures framed and flickering against the fade of evening's light.

As we drove towards Sinéad's, the road stretched out ahead, familiar and unhurried. Beside me, Éireann sat quietly, her smile soft, the promise of another cosy stop before night crept in.

We arrived at Sinéad's Pub as the day began to fold time toward night, the amber glow from the windows spilling across the damp cobblestones like a warm exhale. Around the back of the building, Maverick waited, its paint catching the soft light. There was a quiet pride in the way I began sliding open the door, revealing the little world within.

'Wow, this is lush, but homey, D,' Éireann said, her eyes widening as she stepped closer.

'Yup,' I said with a smile, ushering her in. 'Brit went all in, and I added my touch. Maverick's our pride and joy.'

I guided her through the snug little space, pointing out the wee gas oven and cooktop, the cleverly tucked-away storage nooks, and the big fold-out bed that somehow looked both practical and inviting. 'We lashed out where it mattered and got creative elsewhere.'

I could see it in her eyes, the way she ran her hand over the wooden countertop—Éireann was already at home, safe and settled in her soul. The weight of stress, of landlords and uncertainty, seemed to evaporate in the glow of the van's interior.

'Okay,' I said, sliding the door softly behind us. 'Let's head inside.'

As we walked towards the rear entrance of the pub, I gave her the quick dossier on my sister. 'Sinéad's solid as a rock. Reserved, diligent. She doesn't miss a thing. She's the quiet kind of confident, steady. The pub runs like clockwork under her hand.'

'Ok,' Éireann replied, checking herself over as though she were heading into a job interview. I smiled; she didn't need to impress anyone. Sinéad would have her pegged in an instant.

Inside, the pub was alive in that timeless way old places are. The murmur of conversation mingled with the clink of glasses, and the warm air carried the faint scent of hops and wood polish. Sinéad stood behind the bar, mid-pint pour, finishing a conversation with one of the regulars. When her gaze landed on me, her face softened, but she didn't rush; she never did. That was her way: present, deliberate, always a step ahead.

When her work allowed it, she crossed the room towards us, her movements purposeful, cutting through the busyness. She wrapped me in one of her solid hugs, then stepped back, hands firm on my shoulders as she studied my face. There was that nod, her unspoken way of saying I was in good form, that she was proud.

'Good to see you, D,' she said, her voice steady and warm.

Then she turned to Éireann, pulling her into a hug with the same strength but an added gentleness. Sinéad's eyes appraised her quietly, but the smile that followed was all reassurance. 'I hear you like a good graft,' she said. 'Well, you've come to the right place.'

Éireann smiled back, a little shy but genuine. 'I'll do my best.'

The pub was the place that lived in stories, dark wood, golden light, and an air of comfortable permanence. It was filling up now, the evening crowd finding their usual places. A group came through the door, and I caught Sinéad's eye.

'Looks like rush hour's starting,' I said.

She nodded briskly while shifting back into her role. 'Aye, better get to it.' But first, she handed Éireann a key. 'Rear door's yours now. The van's all stocked up, fridge and all... D, show her about, eh! I'll pop out and check on you later, Éireann, once things settle here.'

Éireann's reply was soft but bright. 'Grand.'

I took her on a quick tour, showing her the essentials before we circled back to Maverick. Inside, I gave her a rundown of how everything worked, double-checking the cupboards and fridge, though I knew Sinéad had thought of everything. It was all there, more than enough.

'Well,' I said finally, my voice dipping, 'I'll leave you to settle in, sweetie. Puddles and I have to hit the road.' There was a pang in our departure; I'd grown fond of her company, and I could feel the same in the long hug she gave me.

'D, I don't know what to say,' she began, her voice cracking on the edges, tears welling up as she tried to find the words. 'All this, being there for me, it, it...'

I placed a hand on her shoulder, steady. 'Look now,' I said softly, 'no tears. I've been fortunate in my life, and for that. I'm grateful. So, seeing you here, safe and forward-facing, that's how it's meant to be. Feels like we climbed a mountain together, eh?'

She gently nodded, a small smile breaking through as she wiped her eyes.

I made my way back to the car, Puddles greeting me with a faint meow, nestled in his pack. Before I started the engine, I turned for one last wave. Éireann stood framed in the van's downlights, her silhouette soft that brought me comfort, her smile a quiet promise. And then I set out with the road stretching ahead, and began the journey home.

After an hour's drive, I pulled over to let Puddles conduct his serious-faced business. He circled with the precision of an old-world inspector, examining every patch of ground as though preparing a report, while I leaned against the car and stretched out the kinks in my back. It was quiet out there, the kind of silence that lets you hear your own thoughts, most of which drifted towards home.

Back on the road, it wasn't long before the cottage came into view, greeting me with its familiar stone-white glow under the night sky. It felt like a gentle smile you hadn't seen in a while, familiar and warm.

It had been a long day, and I wasn't about to drag everything in tonight. I grabbed the necessities, a bag slung over one shoulder, Puddles tucked under my arm like an unruly loaf of bread, and made my way to the door. The gravel crunched underfoot as I waddled inside in one haul, while Puddles let out a small, soft meow that hung somewhere between complaint and comfort.

The old door groaned as I pushed it open, and a sigh escaped me, unbidden. 'I'm home,' I murmured, half to myself and half to the cottage, that always felt more like an old friend than a structure. It greeted me in return with slight creaks under my feet that whispered a lifetime of memories.

I set Puddles down, freeing him from the carry pack, and he padded along behind me, his little claws tapping softly against the dark wooden floor. The cottage welcomed us with its particular stillness, the faint scent of aged timber and a hint of lavender lingering from my last visit. I dropped the bag near the old armchair, shrugged off my coat, and let the day roll off my shoulders. Tomorrow could handle itself. Tonight, the cottage would hold me.

Feeling the faint stirrings of hunger, I wandered to the kitchen, where the familiar order of my preparations awaited. Before setting off to London, I always ensure the freezer is well-stocked, the pantry lined with jars of preserved veggies, and everything in its rightful place. That way, on the return trip, I can avoid midstream shopping or the hassles of becoming a fully laden baggage handler. Tonight's plan was simple: sourdough bread from the freezer, toasted and heaped with tuna and a flourish of preserved veg.

As I rummaged through the freezer for the bread, my gaze fell on Puddles, sitting serenely in his habitual loaf-like repose. 'Only a small portion of dry food for you, mister,' I murmured, shaking it into his bowl with what I hoped was just enough authority to remind him who's in charge, or at least who likes to think they are.

He made quick work of it, polishing it off in record time. Moments later, as the kettle reached its crescendo, Puddles was back circling my feet, gazing up with wide, pleading eyes that would've earned him a BAFTA for Best Dramatic Performance.

'No,' I replied, my tone as flat as the sea on a windless day, meeting his gaze with the unimpressed look of someone immune to theatrics. 'That might work on Mum, but not me.' He responded with a single, theatrical meow, so I caved slightly, leaning down to give his nose a quick scratch. 'There, that'll have to do.'

The tea brewed as I assembled my modest feast. Toast, golden and inviting, received its crown of tuna and vibrant preserved vegetables. With everything in order, I moved to return the jar to the pantry when I noticed something was amiss.

My plate.

Where the tuna had once sat in abundance, there was now only a sad heap of preserved veg atop the toast. My heart sank, and irritation surged. Either a poltergeist had developed a taste for fish, or Mr Fur-ball had struck. I peered under the table and found him sprawled out, smug and sated, his eyes half-lidded in bliss. He looked like a sun-drenched crocodile, jagged teeth bared in a lazy grin around a mouthful of my pilfered tuna.

'Seriously?' I sighed, exasperation softening into the inevitable resignation. He blinked back slowly, utterly unrepentant. To scold him would be as futile as shouting at the wind. So, with a resigned hmph, I left him to his ill-gotten gains and started over.

This time, I managed to retreat from the kitchen unscathed, toast and tea intact. Sinking into the couch, I allowed the day's weight to slide from my shoulders. TR's message surfaced again in my mind, as it had throughout the day, a persistent undercurrent of thought accompanied by the lingering essence of love that had briefly swept me off my feet, like a tide that had pulled me under for a moment, before setting me gently back onto the shore.

With my late-night snack polished off, I called it a night and shuffled off to bed. My anticipation simmered just below the surface. Tomorrow could wait; tonight, I was eager to finally check that message.

Chapter II

The Opening Between Worlds

The morning moved slow and unhurried as I lay in bed, Puddles curled beside me, his weight a warm certainty, his breath a gentle tide, rising, falling. The hush of the room held steady, the light filtering in, soft and pale, tracing quiet shapes along the walls. With a stretch and a languid reach, I drew my laptop close. There it was. TR's message, waiting like a sealed letter on a quiet desk.

It read:

Hi there D,

Before anything else, just tell me if you felt something when I sent this. Don't describe it. Just say yes if something stirred, moved, or shifted in your space.

Now, about the Zeus dream.

There's a useful point of reference I'd like you to look up: *Carl Sagan on the Human Brain — 360p*. It's brief, but it offers some evolutionary perspective, particularly on the architecture of the brain and the long arc of human development. It's not about interpreting your dream directly; it's more of a lens.

What Zeus represents isn't mystical, necessarily; it's closer to a kind of emergent insight. Think of it as the slow unwinding of chaos into something shaped, some deeper memory of fate arranging itself. The maze, in that sense, is not just confusion, but structure born from confusion- a process. It's the tangle we've made trying to forget we're not lost.

The snake on Zeus's arm is worth pausing on. It's symbolic of what Sagan refers to as the "R-complex," that primal layer of the brain that governs instinct, the reptile brain. In your dream, the snake is moving downward. That's significant. It suggests not suppression, but grounding. A descent into stability. Integration, just putting instinct back where it belongs. Letting higher cognition breathe.

This isn't meant to be overly complicated. What we're circling here is simple: humanity suffers from a kind of evolutionary inertia, a type of sickness, a failure to fully employ the neocortex or the rational self. In that sense, Zeus becomes a metaphor, not a God but a force of coherence, of wisdom reclaiming its role over instinct. We'll revisit all of this as things develop. For now, just ponder over it.

Truth and Love,

TR

I watched the video, letting TR's words settle into a chorus with Carl's. Leaning back into the pillows, I squinted at the ceiling, as if somewhere in its blank expanse, the pieces might arrange themselves into something clear. And slowly, they did. The world's feverish pulse, this mass madness, how else could it be seen but as an illness? A sickness of the mind, of instinct left untethered. It defied logic and reason, defied all sense, just as TR had said.

I didn't pause for long. Fingers to keyboard, I typed my reply:

Hi TR,

Yes. I did experience something, something profound, right around the time your message arrived. A shift, a presence. I felt it strongly.

I watched Carl and let his analysis settle, and the video lingered with me. Your reading of the dream and the connection, there's something in it, something that rings true.

I'm intrigued to hear more.

Kind regards,

D

Satisfied, I logged out, closed my laptop, and lay there. How did he know I'd experienced something at that precise moment? And why hadn't he asked for details about what I'd felt?

Before I could slip too deep into questions I wasn't ready to answer, something steadied me, a quiet reassurance, like a presence just out of sight. Not a voice, not words, just a feeling that whatever this was, I didn't have to hold it alone.

As I drifted off to sleep, my eyes opened, though only in the dreaming sense. I was on horseback, the warmth of the animal beneath me solid and certain. A man rode in front of me, and I held on, the movement of the horse rocking us forward through a canyon lined with stone monoliths, ancient but alive in a way I couldn't explain. The place felt like a memory, but not mine. Something older. Something forgotten. The kind of architecture that shouldn't be possible, but there it was, etched against the sky.

The horse moved steadily, its hooves a soothing cadence. I held on and drifted further; I let myself sink into its rhythm, further and further, until there was nothing else.

I awoke to the stillness of the morning air. No horns, no voices, a quiet bliss far removed from the hum and bustle of London. The only sound was the low rumble of my non-mechanical, purring feline alarm clock. Puddles stretched out beside me, let out a soft trill, then nudged his head under my chin, angling for attention with the ease of one who knew he'd win.

A strange certainty tugged at me, a sense that TR had replied. With a lazy stretch, I reached for my laptop on the bedside table, sliding it onto my lap as I propped myself up against the pillows. Puddles, ever the opportunist, seized the moment to nuzzle under my chin, his morning affection routine now in full swing.

A strange certainty tugged at me, subtle, but certain. TR had replied. I knew it before I reached for my laptop, before I even logged in. Just a feeling, like a thread pulling at the edge of my awareness.

I slid the laptop onto my lap, one hand absentmindedly scratching Puddles behind the ears as he leaned into me with the weight of a creature who had no doubt he'd get what he wanted. His purr deepened, settling into my ribs as I clicked through.

There it was, a new message from TR.

It read:

Hi D,

Good, you felt it. That's the thread. That's where it begins.

Normally, I'd ask for permission first. But... we don't quite have the luxury of time right now. Sometimes, alignment just needs to happen when it can.

If you're open to experiencing it again, could you take note of the exact GMT time when it occurs? Nothing elaborate, just a few words. A sense. Any perceptual shift, emotional movement, or physical change. Whatever stands out.

Give it three days.

I'll line up what I've got and see what speaks back. I should have enough to cross-reference, enough to offer something a little clearer. A deeper picture, maybe.

I know, it sounds clinical. And in a way, it is. But at least for now, I need to approach this from the side of observation. Start with the science... the rest will follow.

Truth and Love,

TR

Smiling faintly, I typed my reply:

Hi TR,

Yes. I most definitely experienced something. Lay it on me any time..lol. I'll certainly make notes and track the times.

Kind regards,

D

I lingered for a moment, caught somewhere between curiosity and the quiet hum of something unique unfolding. A feeling, not quite expectation, not quite certainty, but something close.

Then, like clockwork, TR's status flickered to life. Logged in. A brief moment, a presence just there on the other side of the screen.

And then, gone. A message left behind, a breadcrumb in the dark:

Ok, D,

All is well. I'll write to you in three days, and we'll go from there.

Truth and Love, TR

And then, as I closed the laptop, beyond the soft click of the lids, a quiet punctuation. A warmth spread through me, from somewhere deeper, like the embrace of something unseen. It wasn't the words themselves, not really. It was more... the feeling. The sense that something was whole in myself, like a puzzle piece clicking into place without noticing.

Anna's voice echoed in my mind, nudging at the edges of my subconscious, resurfacing, faint but certain. She'd said it, in the depth of her words. TR's not like anyone else. This wave of Love and reassurance, arriving with perfect timing, seemed to confirm her conviction. It wasn't just a matter of easing doubts, was it? No, it was something more, something like an experiment, but not the kind you could measure with a ruler. More like... a test of something beyond the usual, beyond logic.

Shaking off my thoughts, I swung my legs out of bed and set about my day. Spring had worked its magic, though its charm had decidedly overgrown with vigour. The weeds in my veggie patches had sprung up with malicious energy, while the garden beds cried out for attention. And the lawn, good heavens, had grown into an unruly, towering wilderness that promised to test both my patience and the limits of my lawnmower.

After a quick breakfast, I gathered my tools, then rolling up my sleeves with a sense of purpose. I embraced the challenge rather than yielding to any weight of burden. The garden awaited, a wilderness of weeds and overgrown grass, but as I stood surveying the unruly patch of earth, a quiet thought stirred within me. This day wasn't merely about taming the wild; it carried the subtle weight of something deeper, something gently unfurling, like a seed taking root beneath the soil.

After about two hours of wrestling with the weeds, some so tenaciously deep-rooted that they demanded the heft of a large fork to unearth, I finally stood back to survey my progress. Puddles had been my intermittent supervisor, weaving around me in curious arcs, occasionally swatting at loose blades of grass or nuzzling my sides. Every so often, I paused to catch my breath and headed for some cool water from the kitchen fridge. But now, with the garden beds and veggie patches looking contented and orderly, I took a deep breath and exhaled.

'Done,' I whispered to no one in particular, though Puddles flicked his tail in acknowledgement from his perch nearby.

Then it happened. A deep wave of Love, like before, washed over me. But this time, it was more profound, its force leaving me breathless. I sank onto the long grass, untamed and wild compared to the neatness I had just achieved, and then lay back, surrendering to the moment.

It felt as though something ancient and instinctual had awakened inside me, pulling me inward and upward all at once. The sensation was raw and overwhelming, yet it felt like coming home to a truth I had always known but never quite touched. My senses sharpened. Every sound, every scent, every blade of grass around me seemed etched in a crystalline depth of clarity as if the world itself had aligned with this moment, or I had aligned with the depth of existence.

I closed my eyes as the sensation flowed, spreading through me like a warm current. It rose into my chest, blooming there until it felt as though my heart might burst. Each beat rippled outward, warmth spilling into my limbs, washing across my skin, then evaporating into the spring air around me. It poured out into the garden, mingling with the breeze and the earth, yet it pulled me inward again, a call from the Universe itself, as though it were drawing me toward a divine truth that transcended time and space.

When it began to recede, it left behind a resonance, as a lingering chord played in a vast cathedral. Slowly, I opened my eyes; the stillness broken only by Puddles climbing onto my chest, his purring a low, grounding vibration. He looked into my eyes with a calm assurance, as if welcoming me back from wherever I had been. I hugged him gently, feeling as though a boundary between his being and mine had unfurled, spilling over in a way that words could only clumsily attempt to describe.

TR. The thought surfaced, clear and insistent. I glanced at my watch and made a note of the exact time before heading inside to my journal.

I sat at the kitchen table, trying to pour the experience into words. The pen scratched against the paper, but no matter how I tried, the essence of it resisted being confined to the page. Still, I made note of everything, the time, and that Love, so deep and wide, that swept over me like an endless tide, touching every part of my soul.

When I was done, I lingered with the essence of it still pulsing faintly within me. It felt wrong to rush back into the day, so I made myself a good cup of tea and carried it out to a shaded garden chair. I sat there, letting the breeze play across my skin, bathing in a sense of beauty so profound it seemed hidden, in plain sight within the natural world. Everything felt interconnected, part of a deeper design I could almost perceive, just out of reach.

The garden swayed gently in the spring air, and for the first time in a truly profound way, I allowed myself to simply be, the depths of myself.

I lingered for an hour, perhaps more, reflective and immersed in the ebb and flow of being, just being, as though time itself had dissolved, leaving only the whispering wind, its gentle pluck and pull among the leaves, and the grass stirring into ripples, in an endless, unbroken melody of life. Puddles lay nearby, his soft rhythm of breath rising and falling, anchoring me in the moment. At last, I stirred, my limbs stiff, reluctant to leave the serenity behind, yet, I had to crack on.

'Alright, D, time to shear this lawn,' I declared, the sound of my voice punctuating the stillness. like a hefty stone thrown into a placid lake. The garden called to me, a duty, a tether to the practicalities of life. More than that, I sought the grounding satisfaction of being in motion, ready and open to whatever the ether might send my way.

The mower resisted, its stubborn will as fierce as the weeds that had claimed the yard. I pushed, pulled, and wrestled its reluctant growl across the unruly grass. When, finally, at last, the final patch was done, I paused, the tangy scent of freshly cut grass lingering around me, a natural perfume, thick and satisfying, marking the end of the task. My stomach rumbled audibly, as if in reply. I wandered inside and began rummaging through the fridge and pantry until I'd cobbled together a feast that practically overflowed with goodness.

Puddles appeared, padding over with his rounded, far-too-comfortable gait, his eyes wide with an expectation born of habit. "No, you've had enough," I told him firmly, though he voiced his discontent with a theatrical meow. I relented, giving him a modest portion while ignoring his melodramatic protest, and decided we both needed a stroll to balance out the feast. The sun hung low in the sky as we wandered down the way, its golden light softening the edges of the world. Puddles, true to form, took frequent pit stops, examining his surroundings with an air of great importance before lounging in patches of grass.

During one of his pauses, it struck again, that wave. A Love so vast, so all-encompassing, it seemed to wrap itself around me and within me, a gentle tide compared to the earlier deluge, but no less profound. I lowered myself to the ground, where I settled at the base of an old tree, whose roots seemed to embrace me, cradling me as though I belonged there. The sensation coursed through me, and I made a mental note of the time, as TR had instructed, storing it like one might press a flower between the pages of a book.

By the time we returned, dusk had folded itself into the quiet of night. I found myself hungry again, and thirsty too, which struck me as curious. I ate lightly, drank deeply, noting the emerging pattern. It felt as though each wave demanded something of my body, some recalibration I couldn't yet fathom. I scribbled down my observations, the act of recording steadying the swirl of thoughts in quiet reflection.

The air was mild, almost balmy, so I flung open the windows to let nature weave its way through my senses. The night seemed alive with whispers, and I spent the better part of the evening curled up on the couch, Puddles purring softly at my side. There was an ache within me, a love pang, though how I recognised it, I couldn't say. It felt ancient, like a memory unearthed, a song remembered though never learned.

On impulse, I logged onto the forum. A part of me, foolish, hopeful, sought a message from TR. Though I knew it wasn't yet time. Three days, he'd said, and only one had passed. Still, I couldn't entirely resist, and I couldn't silence hope. What struck me, though, was the activity: an uptick of posts on the forum, not just recounting dreams but describing an inner shift, a sense of change. It was varied, yet different, connected by an undercurrent of shared experience.

Perhaps it was my imagination weaving threads together, but I sensed a connection between my day and the murmurs on the forum. Still, my mood leaned more towards quiet observation than interaction, so after logging off, I found myself drifting into the depths of prophetic videos. One in particular "The Mythic Call" struck a chord deep within. Its imagery, blue star embers, surrender to love, the culmination of cultures, and the geometry of the cube, stirred something profound in me. The thunder at the end of the video resonated like a symphony in my soul, and my thoughts leapt instantly to TR.

And then, as though in answer, a forgotten childhood dream surfaced, clear and vivid, as though it had waited all these years for this moment. I was young, standing before a crystal, like a cube, its glow the same celestial blue as in my recent dreams. Beside it stood a great angel, their form hazy yet immense with a presence that filled me with awe. 'May I touch it?' I had asked, my voice small, trembling with wonder. The angel replied, gentle but firm: "In time."

I had set the dream aside back then, thinking it tied to the penannular brooch inlaid with lapis lazuli, my father had given me, a child's imagination running wild. But now, remembering it so vividly, I felt its significance anew. It seemed intertwined with the recent dream of the door, a cosmic, liquid blue too beautiful for words, rippling as though it were alive.

'Steady yourself, D,' I murmured under my breath. The day had been extraordinary enough without letting my imagination run unchecked. I needed rest, to sleep on these things and let my mind and body rejuvenate beneath the gentle weight of slumber. Tomorrow, perhaps, the threads would weave themselves into something clearer.

The following two days unfolded in much the same rhythm, a gentle pattern of pottering about the cottage, immersed in my time with Puddles, my ever-curious feline companion. The hours passed quietly, punctuated by occasional surges of Love, washing over me in profound waves I could scarcely contain. I documented each one, scribbling hurried notes, though words never quite captured the experience. Reflection followed, always, as though each wave demanded I sit with it, understand it, and let it shape me in some ineffable way.

Love strengthened me. There was no denying it. Each day, I felt a kind of earthy contentedness settling in my chest, as though roots were anchoring me more firmly to the earth while my spirit stretched skyward. It wasn't just joy, though there was joy; it wasn't just peace, though that was present too. It was more, a sense of being utterly in tune, as though I had found the right key and was finally playing my life in harmony with the world around me.

It struck me how often people spoke of Love, tossing the word around as though it were a simple thing. But knowing it this way, truly knowing it, was something else entirely. It wasn't a fleeting emotion or a comforting ideal. It was alive, a force that moved through me, filling every corner of my being until I overflowed. My cup wasn't just full; it brimmed over, spilling into everything I touched, everyone I encountered, even in thought. The desire to pour it out into the world was as undeniable as it was humbling.

On the third night, as the dusk shifted into the stillness of evening, I went to bed. The day had traced itself upon me, leaving a quiet fatigue, a gentle ache in my limbs, and a mind lulled into stillness. But as I closed my eyes, I became aware of it again, that blue. A celestial hue, subtle yet pervasive, as though I was bathed in it, not just around me but within me, too. It pulsed softly, like a distant heartbeat, filling me with calm. It felt like a blessing, a quiet reminder of something vast and immeasurable...

'*Grace*,' I whispered into the stillness, the word rising unbidden from somewhere deep within. It felt right, like an ancient memory rekindled. And then, silence. Stillness. I drifted into sleep, the blue hue carrying me, wrapping me in a gentle embrace that felt infinite, eternal.

Puddles was my alarm clock as I surfaced from sleep slowly, like drifting up through layers of warm, dark water. The quiet was thick, but not empty; something held me there, some lingering presence, soft and steady. It wasn't just rest. It was deeper than that, like I'd been carried somewhere far off, and brought back, stronger, and whole.

I sat up, rubbing the sleep from my eyes, with the morning air cool against my skin. And then, just there, it settled in my chest, a deep knowing. Not sharp, not urgent, just a quiet certainty. TR had sent a message. The thought wasn't something I reached for; it had been waiting for me, like the scent of earth after rain, familiar before I'd even named it.

I moved through the morning slow, unhurried, letting the weight of sleep shake itself off in pieces. The kitchen was still cool from the night, the air thick with the promise of heat later, but for now, it was just me, the quiet, and the steady rhythm of habit. Coffee first, something dark, rich, grounding. The scrape of the grinder, the low gurgle of the machine, the first curl of steam rising like a signal from some small, private ritual.

Puddles wove between my ankles, watchful, as if sensing something unspoken in the air. I let him shadow me to the table, setting my laptop down as I wrapped my hands around the warm ceramic. The first sip settled in, spreading slowly, merging with the stillness.

Then, with the faintest press of a key, the quiet shifted. Logged in now, my mind alert, curiosity stretching itself awake. And there it was. TR's name, waiting, steady in my messages.

The words that greeted me were both affirming and quietly extraordinary:

Hi D,

You should have experienced Love in a wave-type form. Also, a cobalt-type blue calming sensation, meaning you might have experienced it in a slightly visual sense. Here are the GMT times: 10 am, 4:30 pm, 11 am, 8:30 pm, 1:30 pm, and 7:30 pm. They should align with what you experienced. Let me know if the times and sensations match up, and we can go over things in more detail.

Truth and Love,

TR

The Love, it was vivid, undeniable, exactly that. And the cobalt blue? It had been as real as the Love itself, a calming hue that seemed to have bathed every fibre of my being the night before.

The times TR listed floated in my mind, and I felt confident they matched. Even so, I reached for my journal, flicking back to the pages where I had hastily documented the surges of Love. And, sure enough, the times corresponded almost exactly, give or take 30 seconds here or there. I knew any discrepancy was my own, a brief hesitation in glancing at my watch or reaching for a pen.

I felt a moment of amazement, but it was an amazement that somehow folded itself into the flow of everything I had been experiencing, a sense of the extraordinary woven seamlessly into the ordinary. How was this possible? I didn't know, and yet it felt entirely natural, marvelling without seeking to understand.

Puddles nuzzled against my leg, grounding me in the moment as I closed my journal with a sense of quiet awe, my mind already formulating an email to TR. There was more to explore here, more to uncover, and I was intrigued and ready.

I finished my coffee, the taste now faint on my tongue, and opened a new window to begin my reply.

I sent TR an email, my fingers moving with a kind of urgency that felt both deliberate and hesitant. I recounted, in painstaking detail, everything I had experienced, the moments vivid, the sensations lingering as if still imprinted on my very being.

And yet, as I typed, a flicker of awkwardness stirred within me, delicate but persistent, as the faintest ripple on a still pond. The profound, otherworldly nature of it all, woven so seamlessly with the revelations since the Zeus dream, seemed to pull me forward, leaving me with a question I could no longer suppress.

At the very end of my email, hesitating only briefly, I wrote:

I know this might sound out there, TR. It might sound crazy or intrusive, but after everything I've experienced so profoundly, are you JC..?

His reply came swiftly, calm, and direct:

No, D,

I'm just me...

This is about alignment, nothing more mysterious than that. Truth has a way of aligning the mind, gently but precisely. It doesn't require belief. It just meets you where you are.

And no, you're not crazy. I've been asked that before. More than once. It's a natural question when the world begins to tilt in a new direction.

What you've experienced... It's not madness. It's contact. Something brushing against the edges of what we call reality, creation's waters, beyond the seams of this dimension.

It's hard to talk about without sounding poetic or unhinged. But that doesn't make it untrue.

The irony, of course, is that the Truth, when it finally shows itself, is stranger, deeper, than fiction.

Love and Truth,

TR

Over the next two days, we exchanged a flurry of emails. I felt as though I'd stumbled upon a hidden doorway to the Universe's secrets. A calm enthusiasm bubbled within me, and with each message, I sensed TR's reassuring presence, with a wholeness that seemed to wrap around my entire being.

His replies were coherent, grounded, and astonishingly perceptive, as we delved into a wide range of subjects. My curiosity led to long, probing emails, and TR responded with clarity, patience, and insight that felt like light refracting through a prism, revealing new depths to every question I had.

And then, the email came. The one that led to this book you're reading now.

TR wrote:

OK, D,

Now that you've got a clearer sense of the dimensional landscape, what we're actually working with, it's time. I need to write this book. Or rather, something like a book. It won't follow a traditional structure. It'll be more of a collage. But then again, so is life.

The stars, metaphorically or otherwise, are aligning. And the gate... is narrow.

I believe you have both the heart and the skill to help bring this into form. And I won't pretend it's just about writing. It's about bypassing the usual machinery, no agents, no publishers, no middlemen. Not just for reasons of time, or privacy, but because the system itself has become an obstacle.

Publishing, as it stands, is an economy of exploitation. I understand the math. And I won't let this be swallowed by it. This book won't belong to a corporate entity. It belongs to the organisation I'm building, and by extension, to the people it's meant to serve.

A good place to begin might be with your vision of Zeus. That felt like an entry point, something archetypal. From there, we trace a path forward toward the Theory of Everything.

And to be clear: the TOE isn't a final declaration. It's not dogma. It's a living theory. It adjusts. It refines. It corrects itself. Where it leads... that's still open. But this feels like the beginning of something real.

As for what lies beyond our dimension, beyond the TOE, yes, that part's still forming. Still sensitive. But we'll get there.

The offer's simple, D. Do you want to help me write this?

Truth and Love,

TR

I read his words and let them settle. For a long moment, I simply stared out of the window. Outside, the world unfurled itself, more beautiful than ever, every detail sharp, yet mysteriously deeper and alive with purpose. There was a depth to my appreciation now, a connection I had never felt before.

In that moment, I felt an overwhelming, undeniable love for TR and a profound love for an "undefinable God," as he had described, a presence vast and unfathomable, gently cradling the fragile contours of my human understanding.

Words felt so small in the face of it all, and yet words were all I had. And the answer? It was simple. How could I say no? Love itself compelled me.

I replied:

Of course, TR,

Though language may falter, and surely it will, I'll try. For what's unfolded over these last few days, what's settled into me like weather, like the hush before rain, is not something easily caught in words. Still, I'll reach.

The clarity you've offered... It's like something ancient remembered, something that had always been waiting beneath the noise. My mind feels steadied, my feet no longer wandering. There is ground beneath me now, real, felt, known.

To help carry this knowledge forward, to offer even a thread of understanding to a world so parched for reason, for hope, that would be more than purpose. It would be grace. So, let us begin.

With love, always,

D

TR replied with his characteristic ease, his words carrying that familiar tone of quiet assurance:

Ok, D, all good.

Here's what I'll do. I'm going to start revisiting everything I've worked on around the Theory of Everything. Scattered as it is, it's all there, just needs to be drawn together. I'll compile it, distil it, and lay it out as a broad, coherent framework. Nothing overcomplicated. No jargon-laden scaffolding. Just the essential structure.

From there, we move in steps. Piece by piece. A few subjects at a time, whatever arises naturally.

If you begin the book, the rest should follow. There's no need to oversteer it. We just stay aligned and let it unfold. Go with the flow.

Truth and Love,

TR

I read his words slowly, letting them settle. The simplicity of it struck me, unassuming, direct, yet carrying a weight that went beyond what was written. TR never dressed things up with unnecessary detail, never clouded meaning with excess. It was all clean lines and quiet certainty, the kind that stays with you long after the words have faded from the page.

Even after I closed the email, the feeling remained. I could picture him, surrounded by papers, thoughts unravelling in quiet constellations, waiting to be threaded together. And now, in some unspoken way, he'd handed me a thread..

I placed my hands on the table, the warmth of the wood grounding me in the moment. There was no sense of pressure, no weight of expectation, just the steady hum of possibility and the quiet certainty that the journey ahead would unfold as it was meant to.

"We'll just go with the flow," he had said. And so we would.

Over the last few days, TR and I had fallen into a rhythm, a kind of steady, thoughtful rally, like an old tennis match played without rush or scoreboard. I'd serve up questions, and he'd return them with precision answers wrapped not in complexity, but in a kind of grounded elegance. Each one nudged something open.

And then there was the cabin.

The one I had dreamed of since I was a girl, cradled on a hill beneath tall timber, its wide bay window steeped in that strange, suspended light that belongs neither to day nor night. I'd always taken it as a metaphor, some flicker from the unconscious, until TR described it in detail. The real place. His sanctuary.

A place he had known long before we met, though somehow, I had seen it before he'd ever arrived.

It felt like a folding of time. A quiet convergence. As if the dream had been waiting for reality to catch up.

And the wolf? The watchful guide who led me there as a child, again and again, without haste or fear? He wrote that perhaps it was an old self, an ancient thread of awareness that moves just beyond the mind's reach, nudging, steering, knowing the way long before I did.

Over those same days, we'd circled around the edges of his Theory, laying the groundwork with patient intent. And though I had no real "Scooby," as the Brits might say, when it came to physics, TR had a way of making it feel less like science, and more like a conversation with the universe, less chalkboard and more whisper.

The idea of delving into the deeper layers no longer filled me with hesitation. It felt like an invitation. A subject I'd once believed belonged to other people, to other minds, suddenly seemed to lean toward me, like a page half-turned, waiting.

I took a slow sip of tea; the warmth settled deep inside. I glanced toward the screen, and as if timed to thought itself, a message arrived from TR.

This was just the beginning.

Chapter 12

The Theory of Everything

TR wrote:

Alright, D, The Theory is laid out, clear, structured, no loose ends worth pulling. I've tightened a few threads, reworked some nuances, but the essence remains: coherent and unambiguous.

The foundational framework is solid. Nothing more to add, nothing worth taking away. Some of the more bureaucratic, rigid-minded gatekeepers of science may resist it, not because it lacks logic, but because it threatens entrenched paradigms. That's fine. The Einstein orthodoxy may treat this as heresy, but dogma has never been a reliable guardian of truth. The content speaks for itself.

This book doesn't present the final, rigorous mathematical model of the TOE. That will come later, likely through scientific papers and collaborative efforts from others willing to explore the terrain. This is the conceptual foundation, the philosophical and physical scaffolding.

The clarity of the framework is what matters now. Its testability, its coherence with observed phenomena, and its resolution of long-standing paradoxes. That's why it's here. Even controlled space experiments offer indirect confirmation: rats exposed to artificial Earth-like gravity still showed signs of biological decline. The Theory of Everything provides a more comprehensive explanation for why.

Drop it into the manuscript as is. We keep moving...

Truth and Love,

TR

When I opened the first draft of the Theory, I was struck by its clarity. The more I read, the more seamlessly it all made sense, as if the dots were finally connecting. It wasn't overwhelmingly laden with jargon, but it was precise and profound.

With every paragraph, I found myself nodding, a growing sense of understanding taking hold. It wasn't just the words on the screen. It was the way everything slotted together, revealing an elegant, almost poetic structure.

Drawn in, I spent hours immersed in the world of physics and cosmology. Video after video fed my curiosity, reinforcing the idea that beneath all the complexity lay a fundamental simplicity. TR's Theory seemed to encapsulate the essence of the Universe perfectly.

But I had to rein myself in. The book wasn't going to write itself, and my task was to weave the TOE into the manuscript, not get lost in the labyrinth of theoretical science.

I replied:

Ok,

The final draft is streamlined, TR. I can relate to it even more so, so I'm confident others will too. Very impressed. I'll add it in.

Love,

D

As I pressed send, a soft thrill stirred beneath the stillness. The shape of the manuscript hovered just beyond the edges of thought, half-formed but breathing. It wasn't simply physics rearranged, nor some clever reshuffling of tired equations. No. It was a kind of awakening. A gentle upheaval. Something that reached deeper than science, into the marrow of things. Into us.

Theory of Everything (TOE) Framework

TR's Gravitational Resonance Theory

Conceptualising 4-Dimensional Space: A Tesseract Model

A Note to the Reader

Most people, understandably, don't spend their days pondering gravitational theory or higher-dimensional space. They're working double shifts, raising families, living their lives. And yet, in the background of modern life, there's a quiet assumption that science has it all figured out.

But here's the truth: we still don't know what gravity is, though some might assume otherwise. We can describe what it does with astonishing precision, but its underlying nature remains obscure. That's not a failure, it's a frontier.

This framework was written with that in mind. It doesn't expect a background in tensor calculus or quantum field theory. It assumes curiosity, and that's enough. We'll skip the jargon: not out of condescension, but out of respect because real clarity doesn't need complexity as camouflage. And yes, this theory unapologetically pushes past the stale metaphors and calcified dogmas that still cling to the walls of academic science. This isn't a demolition of science. It's an extension.

So consider this a walk, not a lecture. We begin not with equations, but with a shape: a tesseract. From there, we'll unfold a new way of seeing space, time, matter, and life.

If the physics feels heavy at times, that's okay. The core idea can be grasped without wading too deep. Imagine a fourth dimension. Not science fiction, but something real. A deeper structure casting itself into our world like a living shadow.

At its centre is a resonant, inward-moving wave: gravity. Not merely bending space, but creating it. Not one force among others, but the underlying coherence, the missing link between general relativity and quantum mechanics. That long-standing divide? It begins to dissolve.

This resonant gravity doesn't reject classical physics. It completes it. And in doing so, it offers a working Theory of Everything.

As for life: under this lens, it's not a biochemical accident. It may be something far more significant, an emergence shaped by gravity itself. Reaching outward, yes, but also inward, as if drawn toward a dimension beyond itself. Not spiritual fluff. A hypothesis. That life may be on a path to realise itself. And, in doing so, become part of something vastly greater.

If this puzzles you at times, good. It's meant to. It spans gravity, quantum mechanics, cosmology, and consciousness. It's not built for easy consumption. But if you stay with it, if you read with curiosity instead of certainty, something deeper may begin to unfold.

Dimensional Blindness - Why We Mistook the 4th for the Infinitesimal

For over a century, physics has misread the fourth dimension; not because it was hidden, but because it was too present. Like an ocean current we move through without seeing; it shapes both space and perception. We assumed space was self-contained. That mistake gave rise to patchwork ideas: dark matter, dark energy, and inflation. The equations of General Relativity are mathematically exquisite, yet they've led us into a litany of assumptions. Gravity isn't a curvature. It's a resonance from a deeper geometry. Quantum mechanics seems strange only because we're measuring shadows instead of the source.

If the tesseract analogy holds, then our 3D world is a projection, thin, reactive. Gravity and quantum effects aren't opposites. They're echoes of four-dimensional structure folding in. Entanglement, duality, and superposition aren't paradoxes. Signs we've been working in too few dimensions.

The fourth dimension isn't abstraction. It's embedded. It shapes galaxies and neurons. The blindness wasn't in the math. It was in us.

This dimensional perspective changes everything. It invites us to abandon our insistence on locality, to reconsider causality, and to entertain the possibility that the 4th dimension is not a speculative abstraction but a subtle presence embedded in our lived experience, folding inward and outward, shaping galaxies and neurons alike.

From Stillness to Pattern: Life Touching the Fourth

Primordial Constraints Within 3D Space At the origin of the universe, raw matter and energy emerged but remained confined to three dimensions. Lacking sufficient complexity or coherence, they could not breach into higher-dimensional space. They stayed inert, bound to a surface reality, governed by simple quantum and gravitational rules.

Quantum Complexity and Life's Dimensional Reach

Life is not inert. It organises, adapts, and coheres across scales. At the quantum level, life exhibits resonance, entanglement, and pattern recognition, suggesting a sensitivity to the fourth dimension. This may not be poetic. It may be literal. Unlike lifeless matter, biological systems could engage higher-dimensional forces through their internal quantum architecture.

Black Holes as Resonant Cores

Black holes, once seen as ends, are reframed as resonant cores: regions where gravitational light and quantum fields compress and stabilise. Not singularities, but thresholds. Organising matter through inward force, they become functional structures, not anomalies.

Life as a Dimensional Interface

Life, by contrast, resonates outward. It organises against entropy, embeds memory and reflects awareness. In this model, life is not merely biochemical. It is a dimensional interface, capable of shaping coherence, and perhaps even consciousness, through engagement with the fourth dimension.

The Shape Behind the Mirror: A Unifying Framework

The tesseract, a geometric analogue of the fourth dimension, is not just symbolic. It offers a structural lens through which inward and outward forces can be understood as intertwined across space and time.

Outward and Inward

Forces Cosmic expansion and gravitational collapse are not opposites but harmonic counterparts. Outward force stretches space; inward force compacts it. Both are governed by 4D resonance fields. This resonance shapes everything, from star formation to subatomic behaviour. Expansion and compression are not purely thermodynamic. They are vibrational phenomena.

Quantum Mechanics and Dimensional Overlap

Wave-particle duality, entanglement, and quantum unpredictability begin to resolve under this framework. What appears random may be structured patterns unfolding in higher-dimensional space. Particles behave like waves because they are projections, slices of complex four-dimensional oscillations.

We do not witness the whole. We see cross-sections, intersections between 4D fields and limited 3D perception. Quantum events seem strange because we're glimpsing fragments of a deeper coherence.

Why This Perspective Matters

Unification

This framework bridges quantum theory and relativity, not through patchwork equations, but with conceptual clarity. The tesseract and 4D resonance offer a foundation that connects gravity, quantum behaviour, and biological complexity.

Clarity in Mystery

Phenomena once labelled inexplicable: wave-particle duality, entanglement, black holes, cosmic acceleration- may not be mysteries at all. They appear fragmented only because we're trapped inside the fragment.

Observational Pathways

Instruments like LIGO detect gravitational waves from massive collisions. However, the subtler, continuous gravitational resonances described here - whispers from the fourth dimension - require a new kind of detector. Not tuned to thunder, but to song.

Dimensional Perception

We are biological flatlanders, not because of ignorance, but because evolution has built us to survive on surfaces, not to comprehend deeper space. The fourth dimension doesn't evade us because it's distant; it's foundational. Saturating. Like water to a fish. It curves not only geometry but cognition. It disguises itself as a paradox, slips past our instruments, and bends both quantum fields and spacetime from within.

Misread Signals

Perhaps our models fail not from lack of data, but category error. Dark matter, dark energy, inflation: what if these are projections gone awry? Not missing pieces, but misinterpreted shadows. Space might not be a container, but a resonance event: gravity, not a distortion, but a signal.

A Different Lens

Once we stop imposing and start listening, order emerges. The chaos clarifies and patterns form. What once seemed inexplicable begins to feel composed, not random, not paradoxical, but dimensional.

What LIGO Might Have Missed

LIGO's detection of gravitational waves was a landmark moment. But through the lens of gravitational resonance, it warrants reinterpretation. General relativity sees these signals as spacetime distortions caused by massive bodies. The TOE framework suggests something deeper: 4D resonant waves pressing into 3D space, not warping it but compressing it inward.

If this is correct, what LIGO measures may not be gravitational waves in the classical sense, but rather quantum-resonant effects—subtle echoes of higher-dimensional interactions, glitches of which LIGO registers dozens daily, which are often discarded as noise. Yet some may hold overlooked clues, symptoms of a phenomenon that current frameworks weren't designed to detect.

While squeezing techniques reduce quantum noise, they do not remove ambiguity. Nor do they clarify whether the observed oscillations stem from relativistic events or something more fundamental, such as the continuous inward pulse of a resonant gravity field.

Confirmation bias remains a challenge. Observed signals are filtered through general relativity's expectations. But if gravity is resonant and continuous rather than event-driven and punctuated, our instruments are tuned to the wrong scale. Low-frequency background waves, representing subtle fluctuations within spacetime, remain largely undetectable and unexplored. TOE proposes that space is not passively warped, but actively shaped by resonance. LIGO may be picking up the effects of gravitational interaction, not its cause. Until detection technology evolves and independent methods emerge, we must keep interpretation open. What we've called waves might be signatures not of spacetime's curvature, but of its deeper geometry.

Where Quantum Touches the Fourth

Quantum mechanics, long considered strange, may simply reflect our limited view of a deeper structure. In a 4D resonant framework, its mysteries become more intelligible.

Wave-Particle Duality

Particles don't shift between wave and object. Instead, they behave like standing waves shaped by a fourth-dimensional field. What we observe as "collapse" is the intersection of resonance with 3D perception.

Entanglement

The link between entangled particles may reflect a single wave structure extending through a higher dimension. What appears to be instantaneous action at a distance is simply coherence, manifesting beyond our spatial frame.

Compression, Not Curvature

In this model, space is not warped but compressed. Gravity is a rhythmic inward pull from the fourth dimension, intensified by mass. It shapes quantum space from the outside, not within. The more mass, the stronger the resonance, the deeper the compression.

Gravitational lensing, then, is not the bending of empty space, but the reorganisation of quantum structure under resonant force. Galaxies, filaments, and voids reflect this same inward ordering.

A Unified Field

Fourth-dimensional gravitational light waves do more than curve trajectories. They synchronise the quantum and the cosmic. Resonance unites expansion and collapse, randomness and pattern, particles and galaxies, revealing not chaos, but design.

Why This Matters: A Unified Model

Reframing 3D space as a projection of a four-dimensional structure allows gravity, quantum mechanics, and cosmic form to converge under a single conceptual roof. In this model, 4D gravitational resonance provides the missing coherence, revealing a pattern where once there was a paradox.

Dimensional Compression

Rather than warped geometry, this view sees space as dynamically compressed by fourth-dimensional forces. Resonance replaces curvature, offering a more intuitive mechanism that still aligns with observed gravitational effects.

1. Quantum Phenomena Reimagined

Wave-particle duality, entanglement, and non-locality become signatures of 4D interaction-structured, not stochastic. The probabilistic veil may conceal a deeper, resonant geometry.

1. Dimensional Overlap

Just as a flatland being would misread the contours of 3D objects, we may be glimpsing fragments of higher-dimensional processes. Quantum weirdness may simply be a translation error between dimensions.

2. Towards Empirical Clarity

Current detectors like LIGO catch loud signals--collisions, bursts. But the quiet hum of gravitational resonance remains unheard. Future instruments, tuned to the subtler frequencies of space itself, may detect the background rhythm of the fourth dimension.

Conclusion

The Tesseract model doesn't discard physics; it deepens it. It proposes a universe not curved, but compressed; shaped by resonance, not abstraction. What we've taken for a paradox may simply be structure viewed from the wrong dimension. It's not that the universe is strange. It's that we've been looking at it sideways.

Light in Three-Dimensional Space: Resonant Signatures

Light, in its many interactions with matter, reveals more than optical phenomena. Across disciplines, it exposes resonance as a deeper structural principle; one not limited to electromagnetism, but possibly echoed through gravitational and quantum fields. These examples offer a glimpse into a broader coherence:

1. Phonon Lasers: Coherence from Vibration

Phonon lasers, acoustic counterparts to their optical relatives, produce order through resonance. What emerges is not just sound, but patterned coherence, suggesting a universal principle. If gravitational resonance behaves similarly, it may act as an organising force across both quantum states and large-scale structures. Like a tuning fork embedded in spacetime.

2. Surface Plasmon Resonance: Boundaries as Amplifiers

SPR arises when light strikes metal, exciting electron oscillations highly sensitive to their surroundings. This interaction, precise and structured, hints at boundary-based resonance. If light can do this on a metallic surface, could gravitational resonance impose similar precision at the edge of quantum structures?

3. Optical Cavities: Light Refined by Confinement

Within mirrors, light forms standing waves, modes shaped by geometry. Lasers depend on this. Could space itself act as a cavity, shaped by gravitational resonance? In this frame, quantum coherence may not arise in isolation, but as a resonance-bound phenomenon governed by deeper spacetime dynamics.

4. Antennas: Matter Tuned to Frequency

Antennas respond to specific frequencies by design. What if mass behaves the same way? The TOE suggests matter may "tune" to gravitational resonance just as antennas align with electromagnetic signals. Stability, in this sense, isn't merely structural; it's harmonic.

5. Atomic Resonance: Light as Interpreter

Atoms absorb or emit light only when it resonates with their quantum transitions. This selectivity underpins spectroscopy and defines how we explore matter. But the TOE posits this resonance is not limited to electromagnetism. It may be a glimpse into gravitational fields shaping quantum transitions from beneath perception.

6. Whispering Gallery Modes: Echoes in Form

In circular chambers, sound or light travels the perimeter, forming self-sustaining loops. Applied to light, this yields precision technologies. But viewed through TOE, this isn't merely geometry; it's resonance manifesting through form. Cosmic structures may operate by similar principles, shaped not by stretch or pull, but by self-reinforcing resonance patterns.

7. Photon-Phonon Interactions: Light Converses with Matter

When photons scatter off vibrational modes in a medium, subtle energy shifts occur. This isn't just noise, it's dialogue. Brillouin and Raman scattering show light responding to deeper vibrational structure. The TOE extends this idea: gravitational resonance might do the same, modulating matter from within, too quiet to detect directly, yet shaping energy landscapes across scales.

The Emergence of Light

In the universe's earliest moments, light had not yet moved. The cosmic microwave background records its eventual liberation. But what if light, as we now observe it, emerged not merely from thermodynamic unfolding, but as a byproduct of higher-dimensional resonance breaking through?

If gravitational waves are not just distortions, but structured light-like oscillations in a fourth dimension, then the release of photons might mark a transition not only in energy, but in dimensional coherence. The visibility of light could be the residue of deeper rhythms, frozen into spacetime as the universe cooled.

The 3rd Dimension as a Reflection of the 4th

If the physics of our observable universe is a projection of deeper-dimensional order, then light's properties: its speed, dual nature, and quantum interactions may not be fundamental in themselves, but emergent expressions of a more complex geometry. The coherence we observe in photonic phenomena, such as plasmon resonance, whispering-gallery modes, and optical cavities, could be echoes of a layered resonance structuring reality from a higher-dimensional source.

Take photon emergence. Conventionally, photons are excitations of the electromagnetic field. But what if they are surface effects - ripples - produced by oscillations within a fourth-dimensional gravitational wave? Light, in this view, isn't primary. It's a visible byproduct of deeper, resonating dynamics that transcend our perceptual framework.

Resonance itself, often treated as a feature of specific systems, may in fact be universal. The coherence seen in lasers, phonon interactions, and standing light waves may not be isolated quirks, but hints of an organising rhythm running through every layer of reality. Light becomes more than an electromagnetic entity; it becomes a cypher, an observable clue to a hidden architecture.

Unified Resonance from 4D to 3D

If the third dimension is a resonant projection of the fourth, then gravity, quantum mechanics, and electromagnetism are not separate forces. They're facets of the same underlying resonance. This could explain the uncanny structural symmetry observed across vastly different scales: from entangled particles to cosmic filaments. These are not coincidences; they are the geometry of resonance playing out in space.

In this view, the architecture of the universe, its origin, its form, its evolution are sculpted by gravitational resonance. Light, matter, and even spacetime arise from this dynamic compression. The strange coherence seen in quantum systems may be the visible consequence of gravitational rhythms flowing inward from the fourth dimension.

4D Resonant Light Waves and Key Implications

Infinite Reach of Gravity

If gravity operates as a resonating 4D light wave, its influence wouldn't fade into insignificance. Like light from distant stars, its signal would persist across vast cosmic distances, an unbroken oscillation rather than a force that merely diminishes.

Interaction with Matter

Just as photons interact with matter, these 4D waves would warp quantum space, guide the formation of galaxies, and influence quantum geometry. Gravitational lensing and quantum lensing might be two expressions of the same inward pressure, depending on scale.

Layered Resonance and Density

Gravitational light waves may vary in frequency and intensity depending on surrounding mass and energy. This helps explain phenomena like the Sun's corona being hotter than its surface, or the

Earth's heat peaking in its outer core. These aren't thermodynamic anomalies; they're expressions of layered resonance peaking outside dense centres.

Galaxies, black holes, and biological systems may follow the same principle. Their complexity and structure emerge not from chaos, but from an ordering resonance; denser zones giving way to layered expressions of energy and form.

Biological Systems as Resonant Structures

Life, too, may be a resonance effect. Biological systems build in layers, from organelles to organisms, each amplifying and refining the coherence of the previous. Evolution isn't random noise; it's a resonance cascade, moving from simplicity to intelligence through structured amplification.

Eternal Resonance

If gravity is a 4D oscillation woven into the fabric of spacetime itself, then its rhythm never fades. It doesn't merely move through spacetime; it composes it. This model reframes gravity as the universe's silent metronome, a persistent structure, not a dissipating force. What we've called expansion or curvature might instead be evidence of an ongoing harmonic structure unfolding from a deeper dimension.

Resonating Resonances - Vibro-Acoustics and Clustering Particles

WithinTOE lies a simple but radical claim: resonance is not a side effect. It is the structure. It is the rhythm beneath the noise. And it isn't a metaphor. It's how the universe organises itself at every scale.

Resonance creates order where randomness might otherwise reign. It synchronises waves, clusters particles, and stabilises forces. In vibro-acoustics, sound arranges matter into intricate patterns. The same logic may apply across the cosmos. Gravitational resonance shaping galaxies as easily as sound shapes sand.

This prompts a deeper question: have we mistaken emergent order for fundamental law? What if many 'laws' of physics are simply the artefacts of self-organising resonance systems; patterns that arise not from divine rulebooks, but from harmonic inevitability?

1. Vibro-Acoustics in Nature

Some rodents navigate using vibro-acoustic cues. Sound shapes their world, literally. Particles cluster under its influence. Nature seems to intuitively leverage resonance, not as ornament, but as architecture.

2. Quantum Clustering

Particles, too, exhibit clustering when exposed to resonant frequencies. Not noise, but pattern. These behaviours mirror cosmic dynamics: quantum systems and galactic structures may both be guided by resonance, not random chance.

3. Cosmic Structure

Galaxies don't form by accident. Their shapes, spirals, and rotational stability suggests deeper symmetries. Perhaps gravitational resonance, not just gravitational pull, sculpts their form and sustains their motion.

4. Zero-Point Oscillations

Even the vacuum isn't silent. Quantum fluctuations ripple through it, even at absolute zero. These aren't random. They're structured. TOE posits that these zero-point vibrations may stem from gravitational resonance itself; an omnipresent hum in the background of being.

5. Structure Formation and Expansion

From the clustering of matter to the large-scale filaments of the cosmos, the same rhythm persists. The universe may not simply be expanding; it may be resonating its motion shaped by the feedback between gravitational light and quantum space.

6. Biological Precision

DNA zips and unzips with uncanny order. Proteins fold with geometric consistency. Photosynthesis follows rhythmic energy transfer. This is not random biochemistry; it's resonance in motion. Life may be a biological echo of quantum harmony.

7. The Golden Ratio

From the spirals of galaxies to the petals of a flower, the Golden Ratio keeps reappearing. This isn't an aesthetic coincidence. It may be resonance expressing itself in the most efficient geometries possible, encoded into matter by gravitational rhythm.

8. A Unifying Pattern

Resonance bridges the micro and the macro: it organises atoms, sustains life, and structures galaxies. TOE proposes that this isn't poetic; it's structural. Life and physics are not separate; they are resonant expressions of the same underlying wave.

Closing Thought This isn't mysticism. It's pattern. If resonance governs everything from phonons to planetary systems, then the architecture of the cosmos isn't upheld by force, but by subtle alignment. This model doesn't abandon established physics; it suggests the rules may themselves be the echoes of a deeper reality.

Inward Anomalies: Particle Clues in a Resonant Universe

The Theory of Everything (TOE) proposes that inward gravitational resonance may help explain anomalies in particle physics and cosmology. These aren't outlier curiosities; they might be the breadcrumbs of a deeper organising principle.

1. Quantum Vacuum Energy

The problem: Quantum field theory predicts vacuum energy far greater than what we observe: the infamous cosmological constant problem.

TOE insight: An inward gravitational resonance could act as a balancing mechanism, tempering the outward expansion and offsetting the apparent excess. The vacuum isn't empty, but dynamically regulated.

2. Baryon Asymmetry

The problem: Matter dominates over antimatter, despite theories predicting symmetry.

TOE insight: Resonant gravitational waves may have disrupted this balance early on, tipping the scales in favour of matter through asymmetric interactions within the quantum field.

3. Strong Force and Confinement

The problem: Quarks are never seen in isolation. They're permanently bound within hadrons.

TOE insight: Inward resonance could amplify confinement, reinforcing the strong force's effects and stabilising particle structures through deeper spatial compression.

4. Muon g-2 Anomaly

The problem: The muon's magnetic moment deviates from Standard Model predictions.

TOE insight: These deviations might result from gravitational resonance subtly shifting particle properties; an unseen hand in the background physics.

5. Higgs Boson and Mass

The problem: The Higgs field accounts for mass, but not its precise values.

TOE insight: Mass may arise not just from the Higgs, but from standing waves formed by gravitational pressure. The Higgs could be a localised ripple in a deeper 4D resonant field.

6. Supersymmetry Gaps

The problem: SUSY particles remain undetected despite strong theoretical motivation.

TOE insight: Phenomena attributed to supersymmetry might be better explained through gravitational resonance and quantum fluctuations. No need for exotic particles if resonance is the root cause.

A Silent Rhythm

We feel gravity, but not the resonance behind it. Like heat felt without seeing infrared, resonance may be ever-present, yet undetectable because there's nothing outside it to compare. It's not absent; it's everywhere.

Our perception relies on contrast. We sense motion through change, sound through fluctuation. If gravitational resonance is stable at our scale, there's no perceptual edge to grasp. We're immersed in it.

Detecting the Invisible

Still, where resonance warps, around black holes, neutron stars, or dense matter. It may leave measurable fingerprints. These deviations could expose the hidden structure beneath gravity, quantum fields, and the expansion of space.

The TOE doesn't invent new particles; it proposes that some of our greatest physical mysteries may be side effects of an inward, organising resonance.

Main framework

1. Gravity as a 4-Dimensional Light Wave with Resonating Effect

In this interpretation, gravity is not merely a force between objects but a four-dimensional resonating light wave shaping the universe. Unlike the classical view of gravity as a direct attraction, this model sees it as a subtle, persistent inward pull that grows with mass and organises matter across all scales. Rather than producing sudden bursts of energy, this gravitational wave continuously shapes the structure of spacetime, guiding the formation of galaxies and the development of life.

The Subtle Nature of Gravitational Resonance

Gravity here is not abrupt or explosive. Even near black holes, its resonance becomes more precise rather than fundamentally different. This ongoing, balanced influence integrates systems across time and space. Its subtlety makes it omnipresent yet almost imperceptible; a quiet architect of structure.

Explanation

Each mass, from a subatomic particle to a star, generates its own gravitational resonance. Larger masses intensify this effect, shaping the surrounding quantum space. This model positions gravity as an interaction between the resonating wave and the lower-dimensional quantum fabric we perceive as spacetime. The more massive the object, the stronger the resonance, producing pronounced effects locally and cosmically.

Layering Effect and Natural Sorting

This framework explains natural formations, planetary orbits, moon systems, and planetary rings as effects of gravitational layering. Resonant nodes emerge where intensities align, creating stable configurations. Saturn's rings, for instance, form in distinct gaps due to resonance intervals. This organisation scales from galaxies to atoms, suggesting a cosmos shaped by filtering patterns, not chaos.

Reasoning

Rather than isolating gravity as a fundamental force, this model unifies it as a result of four-dimensional wave dynamics. As mass increases, the resonance strengthens, pulling space into recognisable gravitational patterns. It even invites a rethink of time: could time be an emergent feature of gravitational resonance? Consider atomic clocks, what if their oscillations respond to gravitational fluctuations, making time itself a product of wave interaction?

Gravitational Attraction as Resonant Frequencies

Gravitational attraction becomes the result of resonant frequencies in quantum space. This space is not passive; it deforms more around larger masses. Each object emits a resonance, larger ones with low-frequency waves that warp space and attract smaller ones. Gravity, then, is not a simple pull but a resonance-driven convergence that integrates quantum and classical models.

Implications for Cosmic Structure and Cyclical Patterns

This model supports the idea of a dynamic, breathing universe. As gravitational resonance weakens over vast time scales, cosmic structures may unravel, suggesting a cycle of formation and dissolution; a gravitational exhale and inhale. This view aligns with bounce or cyclical universe theories, where the cosmos pulses through endless phases.

Dual Nature of Gravity

Gravity, classically seen as a force, was reframed by Einstein as curvature in spacetime. The TOE synthesises both: gravity appears as a force due to the resonant wave's structuring effect, but it is fundamentally a wave phenomenon. It influences matter much like sound creates vibration or light exerts pressure. This inward wave structures the medium itself, producing the illusion of a force.

More Than a Force

Gravity sets the conditions for space and energy to behave. It's not merely acting on things; it defines how things exist and interact. It's a framework, not just a mechanism.

Both "*Force*" and "*Not Force*"- Unified in the Resonant Wave

Gravity plays a dual role: it behaves as a force in the observable universe while fundamentally being a four-dimensional wave. Just as light behaves both as a particle and a wave, gravity may be a projection of a deeper form. This could explain why light's speed is a cosmic limit; not arbitrary, but a product of how space resonates under gravitational influence.

Conclusion

By reimagining gravity as a resonant four-dimensional wave, we bridge the gap between classical mechanics and quantum theory. Gravity becomes a dynamic, structuring rhythm embedded in the geometry of reality, not a force imposed, but a resonance unfolding. It is this resonance that may unify our understanding of light, mass, space, and time.

2. Resonant Quantum Space

Resonant Quantum Space is the universe's foundational substrate-an emergent quantum field shaped and structured by gravitational resonance. Forged in the earliest moments of existence, whether by Big Bang or Bounce, this medium has been continuously refined by gravitational interactions, especially around massive bodies like black holes. It acts as the carrier of resonance itself, linking quantum behaviour with large-scale cosmic structure.

Explanation

Every mass, whether a subatomic particle or a star, generates a resonant gravitational field. In this model, gravity is not a mere pulling force but a wave interacting with a lower-dimensional quantum scaffold. Larger masses generate deeper, slower resonances that distort the quantum field, not through blunt attraction, but through harmonics that shape the fabric of space.

The interaction is not about proximity alone; it is about structure. Space is not passive; it responds. The more massive the object, the stronger the resonance, and the greater the geometric shift. Gravity becomes a continuous feedback loop between matter and the quantum field, a kind of dynamic scaffolding that shapes reality not only locally but across vast distances.

Resonant Quantum Space thus replaces the idea of empty vacuum with something far more alive. A medium shaped by ongoing, subtle distortions that echo across time and dimension. It provides the underlying coherence behind everything from galactic formations to quantum strangeness. In this framework, space itself is a living field, resonant and responsive, the unifying tissue of a cosmos in motion.

3. Hopf Fibration: Geometry of Hidden Dimensions

The Hopf Fibration is more than a mathematical curiosity. It offers a way of seeing how higher-dimensional structures might ripple into the world we perceive. Imagine a three-dimensional sphere (S^3) projected onto a two-dimensional one (S^2), with a circular dimension (S^1) woven through every point. At its core, it's about structure: how complexity in higher dimensions translates into the geometry we observe.

1. Resonant Geometry in the TOE Framework

The TOE proposes that gravity isn't just bending space; it's resonating through a deeper, four-dimensional structure. The Hopf Fibration may model how this resonance manifests: not as force alone, but as structured projection. Matter, energy, and quantum interactions could be differing expressions of the same geometry, cast into three-dimensional space. This isn't just elegant mathematics, it may describe how *reality* actually works.

2. Dimensional Projections and Natural Patterning

In this model, gravitational resonance and quantum effects emerge from four-dimensional interactions mapped into our world. The projection from S^3 to $S^2 + S^1$ offers a metaphor and perhaps a mechanism for how complexity reduces into coherent structure. Fibonacci spirals, chirality, and even quantum entanglement may all echo this geometry. Entanglement could arise from connectivity across higher dimensions, rather than any local mechanism. Chirality and matter's asymmetries may be consequences of how higher forms fold into lower ones. Gravity, as a 4D resonant wave, might propagate through spacetime along the logic described by this projection, embedding order where chaos might otherwise reign. If so, many physical laws are not laws the strictest sense, but constraints imposed by the shape of hidden geometry.

3. Toward a Unified Topology

What makes the Hopf Fibration compelling is not its complexity, but its simplicity: a subtle order beneath surface chaos. TOE doesn't stack abstraction upon abstraction; it suggests that reality is already structured by higher-dimensional logic. The particles we observe, the forces we measure, might be no more than the shadows of richer dimensional processes; shaped by geometry, guided by resonance. This isn't mysticism. It's pattern. If gravity, entanglement, and quantum asymmetries all stem from this structure, then what we see is only a partial view. The Hopf Fibration may be a window into the rest; a quiet map that shows how the hidden order beneath spacetime could look.

4. 99.99% of the Universe is Plasma

That plasma makes up 99.99% of the observable universe isn't a footnote. It's a clue, a bridge between gravity, energy, and structure at all scales.

Plasma is everywhere: the stuff of stars, the filaments between galaxies, the glow of nebulae. We usually see it through electromagnetism, parsed via Maxwell's equations and magnetohydrodynamics. But what if that's only half the story?

In the TOE, plasma isn't just electrically charged matter. It's a responsive medium, finely attuned to gravitational resonance. These resonant waves don't simply bend space; they ripple through it, shaping energy and matter depending on density, configuration, and proximity to mass.

This might explain plasma's strange behaviour. Its motion, its twisting filaments, its uncanny structure; what if those aren't random, but resonant? What if plasma, with its fluid balance between form and energy, is particularly sensitive to the subtle geometry of spacetime?

Consider the Sun's corona, hotter than the surface, defying expectation. Or the plasma jets launched from galactic cores, focused, accelerated, sculpted with a precision electromagnetism alone struggles to explain. In this framework, quantum effects and gravitational resonance may be co-shaping these behaviours, revealing a deeper order behind what appears to be chaos.

Plasma, then, becomes more than matter; it becomes a messenger. A translator between the vast gravitational architecture of the universe and the intimate quantum dance beneath it. Another key that turns and clicks into place.

5. Universal Vibration and Resonance

In the Theory of Everything (TOE), gravity isn't just a force; it's a resonance. A subtle but fundamental vibration that distorts and shapes the quantum fabric of reality. What we call gravitational waves aren't surface-level ripples, but expressions of a deeper oscillatory dialogue between mass and quantum space.

This resonance doesn't merely deform spacetime; it animates it. Quantum space responds like a struck string or tuning fork, vibrating in rhythm with gravity's pressure. These waves aren't passive they're sculptural, formative. They give shape to energy, define mass, and may thread the large-scale structure of the universe together.

It's important here to distinguish:

Vibration is motion. Resonance is response. The latter is amplification; geometry awakened by the right frequency.

In this model, gravity generates more than pull; it produces a kind of rhythm. One that echoes through particles, atoms, stars. It hums beneath the drift of atomic clocks, surfaces in the orbits of moons, and stretches between galaxies in long, slow undulations.

Gravitational waves, then, are not background noise. They're the signature of something deeper. An organising principle, a harmonic structure written into the geometry of space itself.

6. Time and the Limitations of Classical Tools

Time, in classical physics, is treated as a neutral, linear backdrop; a ticking stage on which events unfold. But this perspective begins to crack under the pressure of quantum mechanics and cosmic scale. In the Theory of Everything (TOE), time is not a fundamental property, but an emergent feature. An echo of transformations occurring within quantum space, shaped by inward gravitational resonance.

Atomic clocks don't measure time itself. They measure oscillations, electrons responding to quantum-gravitational dynamics. What we call time is not independent; it's a result of how space, mass, and energy resonate. The fact that atomic clocks tick differently depending on altitude or gravitational intensity isn't just a curiosity; it's evidence. Time bends not because it exists absolutely, but because the resonant structure it emerges from does.

Even Einstein suspected something deeper. In later reflections, and in his conversations with thinkers like Tagore, he hinted that time may be relational-a result of interaction, not an absolute. TOE picks up that thread and weaves it through a new frame: that time is a shadow of resonance, not the container of it.

Classical tools like atomic clocks, LIGO interferometers, and even spacecraft telemetry systems (e.g. the Pioneer anomaly) pick up more than we realise. When they drift, when they return puzzling results, it's not just noise. It's often because they're probing a layer of reality they weren't designed to see. A quantum-gravitational layer, subtly vibrating beneath their assumptions.

Gravitational waves, in this model, aren't just ripples in spacetime. They are pulses through the quantum fabric itself; resonances that ripple through the very substrate from which clocks, matter, and perception emerge. When LIGO detected them, it didn't merely confirm Einstein. It brushed against something deeper: a quantum architecture where space and time are no longer fixed, but responsive. These waves may be the reason oscillatory systems drift. Why even our most precise instruments must constantly recalibrate, as if adjusting to an invisible rhythm.

The speed of light presents another challenge to classical thinking. Photons don't experience time. To them, distance doesn't exist in the way we perceive it. TOE suggests this is no mystery. Photons operate on a different dimensional footing; time and space are emergent features of gravitational resonance, and massless particles sit outside that fold. They aren't defying the rules; they simply belong to a deeper one.

Bell's Theorem, long a thorn in the side of realism, may actually point the way forward. Quantum entanglement, instead of being a spooky violation of locality, might be a direct consequence of gravitational resonance; an invisible wave that connects particles beyond our ordinary dimensions. These aren't paradoxes, but glimpses of a structure far more deeply woven than our tools and models assume.

The double-slit experiment reinforces this. It shows not just quantum strangeness, but alignment. Particles behaving as if they already "know" the resonant structure they're moving through. Observation may collapse the wavefunction, but it may also be the moment when the observer, the particle, and the geometry all align within the same resonant field.

This perspective reframes even quantum computing. Qubits don't just decohere because of environmental noise. They might be collapsing due to gravitational resonance- ripples in quantum space subtly interfering with coherence, entanglement, and interference patterns. In this light, decoherence isn't just an engineering problem; it's a clue. Gravity may be silently shaping what's possible in the quantum world.

The tools we built to measure time, observe matter, and decode reality were shaped by a classical worldview. But now they're returning signals from a deeper structure.

Chapter 13
The Theory of Everything II

7. Entropy as a Result of 4D Resonant Light Waves

The four-dimensional resonant light wave isn't some obscure force. It is a hidden rhythm, quietly shaping the structure of reality. Like sand on a vibrating plate, what appears to be chaos transforms into order. Replace the sand with galaxies or the threads of spacetime, and the analogy still holds.

Mass amplifies this resonance. The denser the object, whether a black hole or a young universe, the more intensely it interacts. That intensity isn't just theoretical. It has consequences. Gravitational resonance may be a direct driver of entropy, influencing how matter evolves and rearranges itself. As resonance increases, the geometry becomes more complex. Patterns sharpen. Symmetries fold into intricacies. Massive systems don't just absorb resonance. They help sculpt it.

Entropy and Cosmic Evolution

There's a feedback loop here. Mass shapes resonance. Resonance deepens complexity. The heaviest systems in the cosmos are also the most structurally sophisticated. That may not be coincidence. It might be baked into the laws of thermodynamics themselves.

Entropy is usually framed as decline. Decay. A drift into disorder. But what if that's only half the story? As entropy increases, structure doesn't necessarily vanish. It multiplies. Galaxies emerge not in spite of entropy but because of it. Resonance doesn't just erode. It builds.

What we see is not a universe winding down, but one refining itself through every phase transition. Early simplicity gives way to spirals and filaments. Then to life. Then to thought.

Resonance and the Architecture of Complexity

Gravitational light waves, if they are resonating inward as TOE suggests, may not be flattening the cosmos into stillness. They may be shaping it from within. Entropy becomes a form of selection. Not everything survives the resonance, but what does is more coherent, more complex.

This isn't noise pretending to be order. It is structure, misread. The fingerprints are everywhere, from solar systems to biological symmetry. In this view, the second law of thermodynamics doesn't describe a fall. It describes a process of emergence. Not the loss of order, but the reconstitution of form. Not decay, but transformation.

8. Thermodynamics: The Quantum–Gravitational Nexus

Thermodynamics often feels like settled science. Heat moves from hot to cold. Energy transforms but doesn't vanish. Systems tend toward disorder. And for most practical purposes, those rules hold. But when quantum behaviour and gravitational fields enter the equation, the framework starts to shift. Quietly at first, then profoundly.

Quantum Entropy and Resonance

Traditionally, entropy is framed statistically. A measure of disorder. A way to track the number of configurations a system can take. But what if entropy isn't just random? What if it follows hidden contours, shaped by something deeper?

If space is not inert but structured, an active quantum field, a resonant lattice, then energy flow may not be entirely chaotic. It may be guided. Not controlled, but nudged by underlying patterns. These wouldn't be visible in classical equations, but they might leave their mark in quantum anomalies or in the strange behaviour of thermal systems under stress.

From this view, entropy becomes less a measure of decay and more a whisper from the architecture of space itself. It reflects how structure unfolds, not just how it falls apart.

Thermodynamics and Resonant Gravity

Gravity, in this framework, is not passive. It interacts with space as resonance, a kind of wave structure that influences local conditions. And when gravity oscillates, as it does near massive bodies, it could subtly shift thermodynamic behaviour.

In extreme environments, such as black holes or neutron stars, we might expect deviations from standard thermodynamic predictions. Not violations, but departures. Equilibrium could become a dynamic state, constantly adapting to local resonant pressures. Heat flow might reroute along gravitational gradients that standard models ignore.

Entropy as Spatial Structure

This brings us to a more radical idea: entropy may not simply describe systems; it may describe space itself. If quantum space is resonant, patterned, and evolving, then entropy might be embedded in its very structure. Not as randomness, but as a directionality. Like grain in wood or tension in a stretched membrane, it could guide how systems evolve without dictating every outcome.

Entropy would then be the signature of quantum space unfolding. It increases not because order fails, but because complexity is being allowed to deepen.

Energy Exchange and Cosmic Expansion

This has implications for the universe at large. If energy moves across a structured, resonant field, then expansion may not be a relic of an early explosion. It may be ongoing, driven by thermodynamic processes within this deeper architecture.

The universe doesn't expand because it must. It expands because its internal structure allows it to. Energy is not merely dispersing into emptiness. It is diffusing across a living, resonant medium.

And that shift, from randomness to resonance, may be the key to uniting thermodynamics with quantum gravity.

9. Quantum Lensing and Gravitational Lensing in Resonant Quantum Space

Gravitational lensing is typically explained as the bending of light caused by massive objects warping spacetime. Quantum lensing, by contrast, operates at a subtler scale: not through mass-induced curvature, but through the behaviours of photons interacting with the resonant quantum field. While one pulls our gaze to the distant cosmos, the other invites us to look closely at the quantum domain itself.

In gravitational lensing, the bending is slow, elegant, and predictable. In quantum lensing, it is immediate, shaped by superposition, entanglement, and interference. In the laboratory, the double-slit experiment hints at these strange distortions. Light is no longer just a traveller along curved geometry, but a participant in a deeper structure that shapes its trajectory from within.

This is where the Theory of Everything (TOE) offers a shift. It doesn't discard classical interpretations but reframes them. Rather than seeing light bent solely by warped spacetime, TOE proposes that gravitational lensing may also emerge from interactions with Resonant Quantum Space. A foundational quantum field subtly shaped by mass and oscillating gravitational waves. Here, lensing becomes not only a geometric outcome but a resonant effect, determined by how light and quantum space influence one another.

Implications for the Nature of Space

TOE sees space not as an inert stage but as a resonant field. Within this view:

1. Gravity is an emergent property of oscillating resonance, shaping both motion and structure.

2. Quantum and gravitational domains are not separate but intimately linked through the fabric of space itself.

By reinterpreting lensing through this lens, we don't simply add to existing models; we reconfigure them. The classical and quantum domains cease to be separate disciplines and begin to converge into one coherent, dynamic whole.

Rethinking the 1919 Eddington Experiment

The famous 1919 solar eclipse experiment, often celebrated as a confirmation of general relativity, might have shown more than light curving around the Sun. TOE suggests an additional layer: the bending may arise from light interacting with the Sun's charged plasmasphere and the resonant distortions of quantum space. As light moves away, the bending diminishes, perhaps not only due to geometric curvature but also due to a fading resonance.

This reinterpretation aligns with another solar mystery: the inexplicably hot corona. Traditional thermodynamics struggles to account for it. TOE suggests that quantum resonance, acting near strong gravitational bodies, might subtly energise plasma in ways that defy classical models.

Einstein Rings and Frequency-Dependent Lensing

Einstein rings, those elegant halos of light around massive bodies, appear to bend unevenly across different wavelengths. That inconsistency raises a question: if all light should curve identically, why the variation? TOE suggests that different frequencies interact differently with Resonant Quantum Space. Light is not just responding to gravity but to the character of space itself: a resonant, fluid-like medium that shifts its influence based on frequency. This isn't just a bending of light by mass, but an intricate dance between resonance and motion.

The Case of Sagittarius A*

Gravitational lensing is expected near supermassive black holes. Yet observations around Sagittarius A* often fall short of those expectations. Classical theory struggles to explain the discrepancy. TOE offers a more adaptable view: quantum lensing is not universal; it depends on local resonant conditions. The relationship between quantum space, mass, and light is dynamic. In this context, the muted lensing effect may not be a failure of general relativity, but evidence of deeper quantum structuring.

Reassessing Supermassive Black Holes

Massive bodies like Sagittarius A*, M87, or Cygnus A are assumed to cause dramatic lensing. But observations often reveal subtle or irregular results. TOE reframes the issue:

1. Quantum Space Compression

Near an SMBH, quantum space may compress into a denser, more tightly structured form. This shift changes how resonance behaves. Instead of sharp bending, light is subtly redirected. It doesn't arc dramatically, but shifts gently, eluding detection.

2. Chaotic Plasma Interactions

The environment surrounding black holes is chaotic, with intense plasma, radiation, and magnetic fields all interacting. These forces distort space, scrambling lensing signatures. Classical theory expects clean curves. What we see instead is complexity. TOE sees that complexity not as noise, but as signal.

3. Light Redistribution

In this model, light isn't simply bent. It's nudged, scattered, and refracted by the resonant quantum field. The result isn't a crisp lensing arc but a more subtle modulation. What seems like a failure of lensing might be a redistribution of light along new, quantum-determined paths.

The M87 Black Hole and Photon Ring

The 2019 Event Horizon Telescope image of M87's black hole offers new clues. The photon ring. A luminous circle around the black hole's shadow may reveal more than classical optics. TOE suggests it reflects quantum resonance at work.

1. Photon Ring Structure

Irregular brightness and subtle fluctuations across wavelengths may indicate interference and resonance effects, not just classical bending.

2. Shadow Asymmetry

The slight deviation in the black hole's silhouette may signal not warped spacetime but a deformed quantum structure, shaped by oscillatory forces at the smallest scale.

3. Plasma Coupling

The turbulent plasma around the ring further complicates the picture. It interacts with the quantum fabric, amplifying fluctuations and producing patterns that defy standard lensing predictions.

Quantum Lensing: From the Cosmos to the Laboratory

Gravitational lensing typically requires the scale of stars or galaxies. Quantum lensing, if it exists as TOE suggests, could be brought into the laboratory. Experiments designed to detect how particles or photons respond to local resonance could reveal lensing effects without the need for telescopes. If these subtle distortions can be observed and mapped, they may shed light on dark matter, unifying optics, gravitation, and quantum geometry.

Implications for the Nature of Space

TOE sees space not as an inert stage but as a resonant field. Within this view:

1. Gravity is an emergent property of oscillating resonance, shaping both motion and structure.

2. Quantum and gravitational domains are not separate but intimately linked through the fabric of space itself.

By reinterpreting lensing through this lens, we don't simply add to existing models; we reconfigure them. The classical and quantum domains cease to be separate disciplines and begin to converge into one coherent, dynamic whole.

10. Acoustic Baryon Oscillations (BAO) and the TOE Perspective

Baryon Acoustic Oscillations are often framed as residual pressure waves echoing through the early universe's plasma. But from the Theory of Everything (TOE) perspective, they may signify something deeper: not just thermodynamic ripples, but gravitational resonance.

In this view, BAO aren't merely sound waves in a hot particle soup: they're gravitational light waves imprinting patterns into a responsive, resonant quantum field. Early space wasn't a passive vacuum, but a pliable medium. Gravity didn't just echo through it; it helped shape it.

Rather than matter bouncing inside a container, TOE suggests that the container itself—quantum space—was part of the formative process. It flexed, resonated, and guided matter into form. These waves didn't just accompany structure; they may have carved its blueprint.

Like sand settling into the troughs of a vibrating plate, the universe's filaments, voids, and galactic clusters may have emerged not from chaos, but from an underlying rhythm. A coherent gravitational cadence that shaped space before matter had even settled.

TOE doesn't discard the standard model; it reframes it. What if BAO aren't just pressure artefacts, but fingerprints of gravitational resonance that transcends scale and connects the quantum with the cosmic?

Here, gravitational light waves aren't passive, they're active organisers. They may have shaped the early universe by resonating through quantum space, laying down the grooves into which matter would later flow.

Resonance, then, isn't an afterthought. It's the organising principle. The patterns we observe today may be echoes of that deeper coherence, one that repeats across scale, from atoms to galaxies.

II. Time Crystals, Oscillating Gravitational Resonance, and Quantum–Biological Emergence

Time crystals are strange. Too strange, at first glance, to be real. And yet, they are. Quantum structures that oscillate without expending energy. No decay. No input. Like a pendulum that never runs down.

In the context of this theory, their existence feels less anomalous and more like a signal. A kind of synchronisation between matter and the subtle pulse of gravitational resonance. Not metaphor, but mechanism. If such resonance pervades space, then time crystals may not merely survive in that field; they may require it. Resonance becomes the groove in which matter rests without disintegration.

This is where physics edges toward biology, not through mystical leaps, but through rhythmic ones. Oscillating gravitational waves are usually cast in grand astrophysical terms: colliding black holes, spiralling neutron stars. But if low-frequency, persistent gravitational oscillations exist, they may nudge the probabilistic dynamics of quantum systems, not with force, but with suggestion. A soft bias beneath the noise.

Biological systems are not strangers to this kind of behaviour. In stochastic resonance, the right kind of noise actually enhances weak signals. It's counterintuitive, but demonstrable. Might gravitational resonance do something similar? Not by driving biochemical change directly, but by tilting the odds: helping fragile coherences hold together where they otherwise might not.

We have already observed that biology behaves oddly in microgravity. Not simply due to the absence of weight, but possibly because an underlying rhythm has been removed. Life may not just endure gravity; it may align with it. Tune to it, like a signal riding a carrier wave.

If so, gravitational resonance becomes more than environmental—it's part of the system's internal logic. Quantum coherence, entanglement, and even consciousness may not be outliers but products of rhythm, synchrony, and alignment.

Time crystals, then, could be more than curiosities. They may be microcosmic reflections of the principle itself- structures that resist decay by embracing oscillation. Not as exceptions to entropy, but as stabilisers within it. Keels on the chaotic sea of quantum behaviour.

And if life evolved in a universe shaped by this resonance, it would make sense that it reflects it. It learns rhythm. Survives by it. Perhaps even thinks through it.

In that light, evolution becomes not just a march through mutation and selection, but a long, quiet dance with gravitational tempo. A species shaped not only by its genes, but by the resonant geometry of the space it lives in.

Time crystals don't need to explain life. But they remind us: coherence isn't always imposed. Sometimes, it emerges because something deeper is pulsing underneath. Something that doesn't push, but invites.

Gravitational resonance might be that invitation. The quiet architect of order.

12. Dark Matter, Dark Energy & Cosmic Expansion

A Quantum Fabric Shaped by Gravitational Resonance

The universe is not an expanding void but a resonant quantum medium, shaped by gravitational dynamics. What we call dark matter and dark energy may not be exotic entities at all, but misread signatures of this deeper interaction.

Light, Gravity, and the Misinterpretation of Anomalies

Take dark matter. Rather than positing hidden mass, the Theory of Everything (TOE) suggests that gravitational anomalies could arise from how light interacts with gravitational resonance, modulations in quantum space that subtly compress or distort spacetime. In this view, the anomalous stability of galaxies results not from invisible halos but from quantum compression maintaining cohesion, as gravity's inward resonance reshapes the local fabric.

Conversely, dark energy, often invoked to explain accelerating expansion, might instead reflect outward dilation: spacetime expanding in response to gravitational waves propagating from galactic compression. These ripples subtly stretch the quantum medium, producing redshifts that mimic acceleration. If gravitational resonance governs both compression and expansion, then cosmic acceleration becomes a natural byproduct, not a mysterious force.

Quantum Lensing and Gravitational Warping

Lensing phenomena, once attributed solely to spacetime curvature around mass, take on a new dimension under the Theory of Everything (TOE). Light bends not just under classical gravity but also due to quantum-scale fluctuations in spacetime itself. These perturbations, shaped by gravitational resonance, can distort light's trajectory in ways that mimic the effects ascribed to dark matter or even compound the apparent expansion rates associated with dark energy.

Black Holes as Quantum Engines

Within this framework, black holes are not merely endpoints of collapse but active agents of quantum space modulation. Their gravitational intensity generates profound ripples that compress and redirect quantum space, shaping galactic structure and contributing to cosmic expansion. Rather than passive consumers of matter, they become transmuters, redistributing quantum energy across spacetime, acting as regulators of balance between contraction and outward flow.

Reframing Expansion: Symmetry, Compression, and Propagation

TOE proposes that cosmic expansion is not an external force but the equal and opposite response to gravitational compression. Galaxies, by contracting space at quantum levels, emit gravitational ripples that push surrounding matter outward. This creates a balanced dynamical system, mirroring Newtonian symmetry: inward pressure, outward dilation.

This expansion isn't uniform. Denser clusters remain gravitationally coherent, merging or interacting locally, while less dense regions expand outward. It's not chaos: it's an emergent structure governed by resonance.

Observational Evidence in a New Light

1. Flat Rotation Curves: Rather than invoking invisible mass, TOE explains galactic stability as a product of gravitational resonance shaping quantum space, coherence without halos.

2. Light and Gravitational Waves: Anomalies in lensing may stem from how light interacts with resonant quantum perturbations, not from hidden matter.

3. Dark Energy Effects: Light travelling through this expanding quantum medium could contribute to observed acceleration; red-shifting not from a separate force, but from changes in the geometry it moves through.

Black Holes and the Quantum Structure of Space

Emerging data, such as from the DESI project, hint at a more complex role for black holes. TOE suggests they are engines of expansion, not mere sinks. If they help generate or reshape the quantum medium through which gravitational resonance travels, they become central, not peripheral, to cosmic evolution.

This reframing does more than reinterpret dark matter and dark energy; it also redefines the nature of the universe. It suggests that we've mistaken symptoms for substances, anomalies for entities.

13. Stellar Black Holes and Supermassive Black Holes (SMBHs)

Stellar black holes keep appearing in the data, heavier than expected, more numerous than our models predict. These aren't marginal anomalies. Some of these collapsed stars register ten times the mass they should, based on what we thought we knew about stellar evolution. If a star of a given size is only supposed to yield a black hole within a certain mass range, then how are these overachievers forming? Is there a missed variable? Some hidden shift in the collapse phase?

Then there are supermassive black holes. Monarchs of gravity sitting at galactic centres. But it's not just their size that raises questions. It's their timing. Some appear early in the universe's timeline, fully formed, already billions of solar masses deep when everything else was still taking shape. By standard models, they shouldn't exist yet. Not unless something is accelerating their growth beyond accretion and mergers.

The gravitational glue holding galaxies together is another long-standing mystery. Outer stars orbit too fast. By Newtonian logic, they should spin off into space. Dark matter was the fix. An invisible mass inserted to patch the equations. Decades on, we've yet to find it. Maybe the anomaly isn't mass that's missing, but a flaw in how we interpret gravity at scale. A deeper resonance might explain what dark matter never has: coherence without a traceable source.

New data from Webb and other surveys reveal galaxies flush with cold gas yet mysteriously quiet, with no star formation and minimal black hole feedback. TOE sees this not as a failure but a feature. With reduced gravitational intensity at the core, there's less compression of quantum space, and thus less energy redistribution. These galaxies are not broken; they're simply less compressed, behaving like softly strummed strings instead of tightly wound systems.

Traditional models explain black holes through collapse, accretion, and collision. But they struggle to account for rapid early growth and the strange distribution of galactic mass. TOE suggests something else is at work. Gravitational resonance, amplifying mass through quantum compression.

Black holes, in this framework, are not just endpoints. They are dynamic systems, interacting with the quantum fabric itself. Under extreme conditions, space may compress in nonlinear ways. If gravitational light waves resonate with quantum fields, then mass may not simply accumulate; it may emerge. Amplified not by brute force, but by rhythm.

This changes the role black holes play. Not passive voids but active architects. Perhaps they shape the geometry of space by translating matter into structure. Silent builders compressing energy into the scaffolding of spacetime. A quiet symmetry, concealed within their darkness.

What if light, nearing a black hole, doesn't just vanish but transforms? The escape velocity argument is clean, but it might be shallow. TOE posits that resonance at the event horizon could trigger a phase shift in the photon's behaviour. Light may not be trapped. It may be rewritten. The event horizon becomes a threshold, not a prison. The border between one physical regime and another.

Physics hasn't seen this yet because the dominant models assume light behaves classically, even in extreme gravity. But what if it doesn't? What if it collapses into a different quantum form near the horizon? That would reframe everything from how we think about energy to the so-called paradox of information loss.

General relativity has been stretched to its edge. It explains motion and light bending well. But under quantum scrutiny, it strains. The ΛCDM model holds for large-scale structure, yet demands that 95% of the universe be made of things we've never observed. That's not a minor blind spot. It's a systemic flaw.

Maybe dark matter and dark energy aren't things. Maybe they're echoes. Artefacts of a resonance we've yet to integrate. If mass is emergent, shaped by quantum compression, then these phenomena start to make more sense without speculation.

The recent discovery of G3425, possibly a black hole in the so-called "mass gap," adds another wrinkle. Too heavy to be a neutron star, too light for a standard black hole. Not an outlier; maybe a clue. The TOE doesn't need exotic particles to explain these cases. It needs a shift in how we understand gravitational interaction at the quantum level.

Black hole mass anomalies and galactic rotation discrepancies mirror each other. They aren't isolated. They are symptoms. The TOE reframes them not as failures but as signals from a deeper order, one grounded in quantum resonance, not invisible scaffolding.

If this holds, then black holes are not endpoints at all. They are central to cosmic structure-shaping, balancing, and expanding space.

And the light we thought they swallowed? It might still be there, just in a different form.

14. Universe Expanding at Different Rates

Recent observations from the James Webb and Hubble space telescopes have confirmed something quietly radical: the universe doesn't appear to be expanding uniformly. Some regions accelerate faster than others.

This challenges the assumptions baked into standard cosmological models, which rely on homogeneous expansion. But within the Theory of Everything (TOE), this unevenness isn't a bug. It's a feature. In this framework, expansion is shaped by gravitational resonance and quantum compression. Regions with more mass, or stronger black hole activity, compress quantum space differently, leading to variations in the rate of expansion. The disparity we observe is not paradoxical, but a natural consequence of mass structure woven into spacetime.

Galaxies Like NGC 1277

Consider NGC 1277, a galaxy that, by conventional expectations, should be thick with dark matter, but isn't. The usual explanations: tidal stripping, anomalous formation history- sound less like understanding and more like patchwork. TOE offers a cleaner account: the absence of dark matter signatures is due to weaker gravitational resonance. Without a substantial black hole or deep mass well, the quantum compression necessary for strong 'dark matter effects' doesn't manifest.

And NGC 1277 is hardly alone. Other ultra-diffuse galaxies, such as NGC 1052-DF2 and DF4, exhibit the same unusual quietness. In standard models, this quiet becomes a mystery; in TOE, it's expected. When resonance weakens, the effects we ascribe to dark matter fade with it.

Rethinking Dark Matter and Dark Energy

Modern cosmology relies heavily on the concepts of dark matter and dark energy. Mathematical prosthetics propping up equations that don't quite balance. TOE takes a different stance. Perhaps these are not independent entities at all, but symptoms of how gravitational resonance structures space.

Where mass is dense, resonance intensifies and compresses quantum space. Where mass is sparse, resonance weakens, and the 'missing matter' isn't missing; it's just uncompressed. Rather than inventing invisible substances, TOE reframes these phenomena as emergent behaviours of a resonant gravitational field embedded in the quantum architecture of space itself.

15. Behaviour of Galaxies: Gravitational Resonance, Quantum Space, and Dynamics

There's something refreshingly honest about stepping back from the tangle of equations and simply watching what galaxies do. Not what our models say they should do, but what they repeatedly show us, even when it makes us uncomfortable.

Galaxies behave in ways that don't quite fit. Some rotate too fast for their visible mass. Others, like NGC 1052-DF2 and DF4, appear to function perfectly well without the dark matter they're supposed to need. Within the standard model, these are glitches to be patched. But through the lens of gravitational resonance and quantum space, as proposed by the Theory of Everything (TOE), they become expected features of a deeper logic.

In the TOE view, galaxies lacking dark matter aren't paradoxes. They may have once hosted active black holes that compressed the surrounding quantum space, producing stabilising effects that persist even after those cores go dormant. The resonance doesn't vanish; it lingers. Like warmth from a long-dead fire, residual compression holds the structure in place.

Now shift to galaxies on the opposite end; dense, fast-spinning, visibly active systems. TOE doesn't call for exotic matter to explain their cohesion. Instead, it points to gravitational resonance: an inward pull shaped by mass and motion, subtly magnified by the influence of central black holes. These aren't just collapsed stars; they're environmental sculptors. They compress the geometry of space around them. And over time, a kind of equilibrium sets in: quantum compression and gravitational resonance align, reinforcing one another like two instruments tuning to the same key. The galaxy becomes a self-stabilising system.

There's a quiet elegance in this view. The structure holds because the space around it has been tuned. It's not that we need to imagine new forms of matter; perhaps we've just misunderstood how the matter we already see is interacting with space itself.

TOE also alters our perspective on cosmic expansion. Galaxies don't merely drift in a passive void; they emit quantum waves that ripple outward. These ripples influence local spacetime behaviour, meaning expansion isn't uniform, but responsive to regional mass and motion. Some areas swell faster, others slower. It's not chaos, it's asymmetrical coherence. The universe, breathing in uneven rhythms.

This dynamic extends further. Galaxies with strong internal quantum compression tend to be more cohesive. Those with weaker compression appear looser, more fragile. TOE suggests that quantum space, its structure and density dictate how tightly a galaxy can bind. That may sound poetic, but it's consistent with what we're seeing in the data.

Even gravitational lensing gets reframed. Traditionally, it's the amount of mass that determines how much light bends. But TOE suggests something deeper: it's the density of quantum space. Denser regions bend light more. Thinner ones, less. It's not just the presence of matter, but the nature of the space it occupies.

Rethinking Dark Matter

Under TOE, the need for dark matter fades. Rather than posit an invisible substance to rescue flawed equations, TOE extends General Relativity by introducing gravitational resonance and quantum compression as the underlying architecture behind galactic behaviour. What we've called "dark matter effects" may instead arise from:

1. Quantum Space Compression: A result of black hole dynamics and gravitational resonance shaping local spacetime.

2. Emergent Galactic Dynamics: Stability arising from self-organising systems, built on the interplay between known forces, not hidden mass.

We don't need to force the universe to fit our expectations. We need to listen more carefully to what it's already telling us.

16. Absolute Rotation

Absolute rotation. Not just movement for its own sake, but a kind of inherent spin, directionality without reference. Even when nothing external is there to compare against, an object still knows it's rotating. And that's more than curious. It's conceptually unsettling.

In the Theory of Everything (TOE), this phenomenon arises from the inward pull of gravitational waves: not the violent ripples from black hole collisions, but gentler, persistent compressions like a slow, cosmic inhale. These waves subtly press inward, shaping the conditions for motion, not as illusion, but as physical necessity.

We see this echoed in precession: the wobble of spinning bodies. It's the asymmetry that grants stability. Imperfection, in this context, isn't disorder; it's structure. The flaws create the balance. Strip those away, and you don't get more harmony; you get collapse.

Stretch this across scale: galaxies revolve, planets spin, even particles exhibit quantum spin, something not quite literal, but still deeply suggestive. Why does the metaphor persist? Because perhaps it's not a metaphor at all. Perhaps it's the same rhythm echoing through different registers of the cosmos.

If TOE holds, rotation isn't incidental. It's embedded. The fourth-dimensional gravitational wave doesn't just bend space; it choreographs behaviour. Rotation becomes a signature, not a side-effect. Spiral structure isn't a coincidence; it's a kind of gravitational handwriting, geometry as resonance.

Consider the paraboloid. Spin two dissimilar fluids in a gravitational field, and the curve appears elegant and predictable. But maybe it's more than an artefact. Maybe it's a clue. A spatial echo of mass, motion, and compression in dialogue.

Natural systems seem to mirror this geometry: spirals, logarithmic growth, the golden ratio surfacing again and again. Coincidence, or signature? TOE leans toward the latter. Gravitational resonance compresses quantum space, folding it inward like gravitational origami. Space isn't passive. It's sculpted. And rotation is one of its core tools.

Take quantum spin: an abstract quantity that lacks classical rotation, yet functions as though it's encoded with directional memory. We accept it because the math works. But TOE asks whether that math is whispering something deeper: that this isn't just abstraction, but an echo of cosmic-scale rotation at the smallest scale.

Even the more accepted terms, frame-dragging and gravitomagnetism, start to feel like patches. Theoretical duct tape. They describe behaviours without explaining their origin. TOE suggests a more foundational coherence: rotation as a geometric response to compression.

Rather than isolated forces, gravity, spin, and resonance emerge as facets of the same structure. Space doesn't just contain rotation; it expresses it.

So what emerges?

1. Structure through motion: Galaxies, stars, and fields are not suspended in space; they're shaped by it. Rotation offers form and balance. It isn't decorative; it's foundational.

2. Geometry as record: Spirals, ratios, curves; these aren't ornamental. They're gravitational signatures. The golden ratio isn't mysticism. It's motion crystallised into shape.

3. Recurrence as meaning: From atoms to galaxies, rotation reappears. It's the common denominator across scale. Eventually, recurrence demands recognition.

4. A bridge between domains: Gravitational waves compress. Quantum space responds. That compression sets the conditions for rotation, mass, and form. This may be where unification begins, not in speculation, but in structure.

5. An opening, not a conclusion: Bringing absolute rotation into TOE isn't a final answer. But it tightens the weave. It adds coherence. It hints that the universe isn't a patchwork of forces, but a choreography.

TOE doesn't claim to have resolved every tension. However, it does offer something rare in modern physics: a sense that the pieces fit.

17. Baryon Asymmetry Problem

It's one of those unsettling facts that sits quietly at the base of our reality: matter exists, but antimatter, its theoretical twin, mostly doesn't. For every galaxy, every atom, every breath, there should have been an equal and opposite counterpart. But there wasn't. Something tipped the scales.

Standard models try to account for this with symmetry-breaking mechanisms, hypothetical particles, and finely tuned conditions. They work on paper. But there's a strange artificiality to them as though we're backfilling an answer into the blueprint, not discovering it.

The Theory of Everything (TOE) offers a different lens. It suggests that this imbalance, this cosmic leaning toward matter, was never a fluke or a glitch. It was an emergent behaviour of quantum space itself: a resonance effect.

In those first moments, when the universe was still forming, space wasn't empty. It was unsettled. Teetering. Fluctuating. But those fluctuations didn't happen in a vacuum. They occurred within a medium already shaped by subtle gravitational waves: low-frequency, directional pulses. Not the dramatic ones emitted by collisions, but something slower. A kind of inward pull from a fourth-dimensional axis.

These waves didn't just ripple through space. They shaped it. Compressed it. And in doing so, they introduced asymmetry, not as an error, but as a consequence of patterned pressure.

Imagine quantum fluctuations guided, nudged, by a directional field: a gravitational resonance that slightly favoured one configuration over another. That leaning, however faint, would compound. Over time, that tilt becomes preference. Matter wins, not because it was selected, but because the environment itself had a bias.

This reframes the problem entirely. We don't need to conjure new particles or exotic interactions to restore balance. We need to recognise that balance may never have been the baseline. The system wasn't neutral. It was responsive. Quantum geometry, under gravitational compression, developed structure with a tilt.

So perhaps the real issue isn't why matter dominated, but why we assumed it shouldn't have.

18. The Resonant Law of Duality: Gravity as the Architect of Balance

Observational Hypothesis Duality is everywhere: in matter and antimatter, in biological symmetry, in cosmic balance. These aren't just recurring motifs; they're structural patterns. This section proposes that such dualities are not philosophical artefacts but physical consequences. Specifically, that they emerge from gravitational resonance: a four-dimensional pressure field shaping quantum space. If this is true, we shouldn't expect chaos, but coherence. Symmetry becomes not an exception, but a law, quietly regulating systems from particles to consciousness.

A Universal Organising Principle Across physical, biological, and cosmic domains, duality presents not merely as a recurring theme but as an emergent property of gravitational resonance: a structural imperative embedded in reality. Gravity, seen here not as curvature but as a four-dimensional resonant wave, acts as more than a background force. It curates equilibrium. It stabilises asymmetries. It does not merely bend space; it balances it.

Atomic and Quantum Duality Particle Stability: The reason protons and neutrons are indivisible isn't arbitrary. It may reflect the discreteness imposed by gravitational resonance. In this view, particle stability becomes a byproduct of space's resistance to fragmentation, a sign that resonance enforces quantum coherence at the foundational level.

Wave-Particle Duality: The photon's ambiguity, sometimes wave, sometimes particle, is often treated as paradox. But under TOE, it's duality by design. Wave and particle are not opposites. They are two coherent expressions of one resonant structure, shaped by a deeper field we're only beginning to grasp.

Biological Duality Symmetry in Life: From the bilateral structure of bodies to the dual hemispheres of the brain, biology mirrors a pattern of balance. This isn't accidental. These mirrored forms may emerge from the same field dynamics that create quantum coherence, as seen in gravity's subtle fingerprint in the architecture of life.

Cognitive Balance: The nervous system functions by constant calibration, with excitation countered by inhibition. This tension isn't chaos. It is stability under stress. A kind of mental homeostasis that parallels atomic structure. If gravitational resonance helps shape subatomic stability, perhaps it shapes conscious systems too.

Cosmic Duality Matter vs Antimatter: The asymmetry between them may not be an imbalance at all, but an outcome of the early universe being pressed, gently but persistently, by gravitational resonance. In this scenario, the universe's 'preference' for matter is a kind of energetic bias embedded in spacetime itself.

Attraction vs Expansion: Gravity pulls. Space expands. Standard models treat this as an unstable equilibrium. TOE reframes it as a resonant dance: expansion, stretching space while gravitational waves apply inward pressure, preserving coherence across scales.

Order and Chaos: Galaxies, stars, and planets form in the midst of chaos, yet fall into curiously ordered structures. TOE suggests this isn't lucky patterning. It is resonance at work, sculpting form through interference and rhythm, much like sound shaping sand into geometry.

Under TOE, baryon asymmetry isn't a mystery begging for mathematical patchwork. It's a physical inevitability born from the resonance of space. Pressure gives rise to structure. And in that structure, a bias. Embedded. Subtle. But enough. Enough to build a universe out of imbalance.

Life as Resonant Duality

Life doesn't merely exhibit duality. It arises from it. Evolution isn't just random variation plus selection. It may also be a resonance-seeking process. Organisms balance genetic instability with environmental stability in ways that mirror the same dynamics governing quantum states. It's not metaphysics. It's physics meeting life halfway.

Gravity as the Architect

Gravitational resonance may underlie:

1. Quantum Scale: Dictating the discreteness and coherence of particles.
2. Biological Systems: Influencing symmetry, balance, and neurological tuning.
3. Cosmic Order: Stabilising vast structures through inward compression and directional coherence.

Gravity, in this picture, doesn't just shape space. It makes structure intelligible.

19. Evolution from Primordial to Complex

From Resonant Waves to Structure

Much like sound waves shaping patterns in sand or light waves interfering to form intricate geometries, the 4D gravitational light wave begins by organising the early cosmos in its most rudimentary forms. But complexity is not imposed; it emerges. Over time, through resonance and interaction with the quantum substrate, this wave may act as a sculptor, coaxing the universe into self-organisation. What begins as primordial simplicity could, through these interactions, evolve into structures of ever-increasing intricacy, much like vibrating surfaces giving rise to elaborate geometric motifs.

Complexity as an Emergent Consequence

As the cosmos matures, so too does its complexity, mirroring how resonant frequencies in different media layer and refine themselves. In the biological realm, single-celled organisms eventually give rise to sentient life through natural selection. Why should the cosmos be any different? The same principle, resonant self-organisation, may underlie the emergence of galaxies, planetary systems, and, ultimately, life itself. Evolution, in this broader context, is not confined to biology. It is written into the very structure of reality.

Languages as Derivatives of a Common Source

Different civilisations develop distinct languages, yet all serve the same fundamental purpose: to interpret reality. This, too, mirrors the nature of physical laws. Mathematics, geometry, and physics may appear as separate disciplines, yet they are merely different dialects describing the same universal interactions.

Unified Yet Diverse

Whether we approach reality through equations, geometric symmetries, or empirical observations, all roads lead back to the same foundational principles, just as human languages, however varied, remain expressions of the same underlying cognition. In this way, the 4D resonant wave acts as a universal syntax, akin to DNA in biology, from which all order arises. The diversity of forms we observe, both in nature and in thought, may simply be different manifestations of this singular, governing resonance.

20. Evolution Driven by Gravitational Resonance

Core Premise Evolution: of stars, galaxies, and living systems is not purely stochastic. It is shaped by gravitational resonance: a fundamental ordering principle that subtly sculpts structure and coherence across cosmic and biological domains.

Gravity as a Sculptor of Complexity Gravitational resonance, particularly its compressive influence on quantum space, helps organise matter into spirals, webs, and nested geometries. The golden ratio, Fibonacci sequences, and fractal symmetries, seen from galactic arms to DNA helices, are not incidental. They may arise from deeper harmonic dynamics embedded in spacetime itself.

Rather than being mere curiosities, these recurring forms suggest that evolution follows not only biochemical and environmental pressures, but also aligns with geometric patterns produced by gravitational resonance. Complexity emerges where matter resonates with structure.

Microtubules and the Quantum Resonance of Life

At the cellular level, microtubules, essential for division, transport, and cognition, may act as quantum resonators. Within the TOE framework, they could subtly amplify gravitational signals, aligning biological processes with quantum fluctuations. This resonance may promote coherence in otherwise noisy systems, enhancing adaptability.

Penrose and Hameroff's Orch-OR theory proposed that microtubules support quantum coherence tied to consciousness. TOE extends this by suggesting that gravitational resonance may help sustain such coherence, allowing consciousness to emerge not from arbitrary neural complexity but from resonant alignment with a deeper structural order.

Coherence as Evolution's Underlying Rhythm

Evolution, then, is more than selection and mutation. It is a coherence-driven process. From the spiral galaxies that form through gravitational tension to the fractal patterns in biological tissues, a unified rhythm persists. Structures emerge not randomly, but through an ongoing negotiation between geometry, energy, and environmental context.

Chloroplasts, for instance, are optimised for absorbing specific light frequencies, hinting at how biological forms tune themselves to local energetic conditions. If gravitational resonance similarly influences matter, it is plausible that cellular structures evolved not just with photonic efficiency in mind, but gravitational harmony as well.

Bridging Biology and Quantum Physics

This framework implies that quantum coherence is not confined to sterile lab environments, but may be an inherent feature of reality, catalysed by gravity's resonant pressure. Structures like microtubules, mitochondria, and synapses may not just operate within classical biology, but sit at the intersection of gravity, light, and quantum information.

In this view, life becomes a medium through which gravitational resonance expresses its complexity. Not as a metaphor, but as a physically resonant phenomenon: biology tuned to cosmic rhythm.

Free Will and Quantum Uncertainty

There is also a philosophical implication. Quantum systems remain in superposition, hovering between possibilities, until interaction collapses them into actuality. That liminal state mirrors the nature of choice. Each biological decision, each adaptation, can be seen as a collapse of potential into form. If organisms operate partially within this probabilistic framework, then "free will" may be a macroscopic echo of quantum indeterminacy.

Under TOE, this is not a mere metaphor. It is structural. Quantum uncertainty, gravitational resonance, and biological flexibility converge into a single coherent model where life evolves through a dynamic balance of freedom and constraint.

Consciousness as Coherent Expression Consciousness, in this light, is not an emergent accident. It is a deepening of coherence. Choanoflagellates exhibit coordinated geometry. Pufferfish create mandalas. Are these instincts, or alignment with environmental and gravitational harmonics?

As complexity grows, so does the capacity for internal resonance. Signals form loops. Feedback emerges. Memory consolidates and eventually, experience arises, not as a mystical event, but as a recursive amplification of coherence across matter.

Consciousness, then, may be what happens when gravitational resonance meets quantum life at just the right scale. Not imposed from above, nor invented by neurons, but grown from the same recursive laws that shape galaxies.

Conclusion

Evolution is not merely a sequence of random mutations filtered by survival. It unfolds within a structured environment shaped by gravitational resonance. The Theory of Everything doesn't reject Darwinian principles; it extends them, offering a broader context in which life adapts not only to external pressures but to an underlying geometric coherence embedded in space itself.

21. Speed of Light

In the Theory of Everything (TOE), light is not merely an electromagnetic wave. It is shaped by gravitational resonance. TOE proposes that gravity itself resonates as a light-like wave, moving at light speed. This resonance sets a natural velocity limit within three-dimensional space and governs how light interacts with quantum space.

Light's Behaviour at the Threshold

According to TOE, when light reaches its upper velocity limit, it begins to express mass in a momentum-based form. Einstein's equation, $E = mc^2$, recognises this, but TOE adds that this apparent mass is not just a relativistic effect. It is a response to the inward pull of gravitational resonance. As light climbs in frequency, its particle-like behaviour strengthens. Gamma rays, for example, exhibit greater mass-like characteristics because they are embedded more deeply in this resonance field.

This implies that light's dual nature--wave and particle--is not simply observational. It is driven by context: the intensity of the gravitational field through which it moves.

Gravitational Resonance and Quantum Space

Light is not gliding through an empty vacuum. It interacts with a structured field. The TOE suggests that quantum space, shaped by inward gravitational waves, provides a resonant framework that calibrates the speed of light. The constancy of light's speed is not arbitrary. It emerges from the consistency of this resonance field, which acts as a background structure for all motion.

In this view, space itself is not passive. It regulates. It sculpts. And light responds accordingly.

Resonant Regulation and Cosmic Order

If the same resonance governs both light and gravity, their interdependence could help explain larger phenomena, such as cosmic expansion. Perhaps the so-called acceleration of the universe is not being driven by a mysterious energy but by variations in gravitational resonance. This subtle field may be guiding light and matter alike, shaping the cosmos through structure rather than force.

Such a view dissolves the boundaries between once-separate concepts. Light, gravity, and space become facets of a single dynamic system.

Quantum Lensing and Speed Modulation

Traditional gravitational lensing bends light around mass. But in a resonant framework, the speed of light itself may also be locally modulated. This is quantum lensing: light does not just bend; it adjusts. Its wave properties shift subtly when passing through dense resonant fields. These modulations might remain hidden under classical models but reveal new structure under the TOE lens.

Conclusion

In TOE, light is not an isolated traveller. It is a participant in a resonant system that links motion, mass, and space itself. This framework not only redefines light's behaviour at high velocities. It points toward a deeper coherence: one where speed, structure, and stability arise from resonance, not randomness.

22. Vacuum Fluctuations: A Constant Phenomenon

In quantum field theory, vacuum fluctuations refer to transient energy shifts that occur even in regions of space considered empty. These fluctuations allow for the spontaneous emergence and annihilation of particle-antiparticle pairs, revealing that the vacuum is far from void. Rooted in the uncertainty principle, this phenomenon reframes emptiness as an active, fluctuating field.

Within the Theory of Everything (TOE), this activity takes on new meaning. Rather than viewing these fluctuations as purely quantum events, TOE suggests they may be influenced and even governed by gravitational light waves resonating through a higher-dimensional framework. These fourth-dimensional gravitational waves do not merely curve space; they permeate it, subtly shaping its structure.

Under this lens, vacuum fluctuations are not random noise. They are the localised expression of gravitational resonance interacting with quantum fields. The ephemeral appearance of virtual particles could thus be seen as a byproduct of deeper resonant tuning within spacetime itself.

This reinterprets the vacuum not as a sea of chance but as a resonant medium, modulated by gravitational waveforms that extend across both the quantum and cosmic scales. Such a framework could bridge the gap between quantum mechanics and general relativity, offering fresh insight into long-standing puzzles like the cosmological constant problem or the quantum behaviour of black holes.

If vacuum fluctuations are indeed shaped by these gravitational fields, we may be glimpsing a deeper coherence behind quantum activity. The vacuum becomes a structured, dynamic field, where gravitational resonance influences the very probabilities that underpin particle formation and decay.

This perspective does not discard quantum field theory; it extends it. By situating vacuum fluctuations within a gravitationally resonant architecture, TOE points toward a more unified vision of spacetime:

active, intelligent, and patterned at every scale. A universe where space isn't empty, and silence isn't still.

Chapter 14
The Theory of Everything III

23. Chirality in Particle Physics

What Is Chirality? Chirality is one of those curious features in particle physics that appears simple at first glance. It refers to a particle's handedness, how its spin aligns with its direction of motion. A left-handed particle spins in the same direction as it moves; a right-handed one spins opposite. While this may sound like an abstract distinction, it has profound implications.

The weak force, unlike gravity or electromagnetism, only interacts with left-handed particles. This asymmetry is known as "parity violation". In essence, the universe appears biased, as if it subtly prefers one spin over the other.

Chirality in the Theory of Everything (TOE) Viewed through the Theory of Everything, chirality takes on a new light. Rather than a static property baked into particles, it may be influenced by gravitational resonance. If gravity resonates as a light-like wave across the quantum fabric of space, could it be shaping chirality itself?

This framework suggests that particles are not inherently left- or right-handed. Instead, they may be nudged into handedness by the resonance patterns within spacetime. The weak force's bias, then, could be a consequence of gravitational influence acting subtly on quantum scales.

Such a possibility reframes chirality as an emergent phenomenon rather than a fundamental trait. What appears as a basic property may actually reflect deeper spatial structures at work.

Unification through Chirality If gravitational resonance underpins chirality, it may serve as a hidden thread connecting the fundamental forces. Weak interactions, known for their left-handed exclusivity, might be responding to gravitational cues embedded in spacetime itself. This could open new avenues in the ongoing effort to unify gravity with the quantum field framework.

Rather than four isolated forces, we may be looking at different expressions of the same resonant principle. Chirality, then, becomes more than a quirk of the weak force. It becomes a marker of deeper coherence.

Experimental Possibilities If chirality is sensitive to gravitational resonance, its behaviour might vary under different gravitational conditions. Testing particle interactions in altered gravitational environments, such as microgravity or near-intense gravitational fields, could reveal subtle shifts in handedness. Such findings would mark a breakthrough, potentially showing that even the spin of a particle is not immune to spacetime's influence.

This also offers new ways to approach the mystery of matter-antimatter asymmetry. If chirality is gravitationally biased, it could be part of the reason matter won out in the early universe.

Conclusion

Incorporating chirality into the TOE enriches both particle physics and gravitational theory. It challenges the notion that handedness is a standalone trait, suggesting instead that it arises from deeper, resonant patterns within space. Chirality may not be a leftover oddity. It may be a clue. A signature of the gravitational waveforms that quietly sculpt even the smallest aspects of our reality.

24. Einstein's Theories

Einstein reshaped the scaffolding of physics. His theories of Special and General Relativity remain foundational, not just for cosmology but for any serious attempt at unification. The Theory of Everything (TOE), as proposed here, stands not apart from his work but as its natural evolution.

1. Special Relativity

Special Relativity reveals that time and space are not static absolutes but flexible dimensions shaped by motion. Its most famous insight—$E = mc^2$—collapses the boundary between mass and energy. Within the TOE framework, this fusion is not just mathematical. It reflects a deeper resonance: a dynamic field connecting energy, matter, and gravity. Mass accrues where resonance intensifies. Gravity is not an external actor but a participant in this transformation.

The Andromeda Paradox

This paradox highlights how even slight differences in motion can lead observers to experience different versions of "now" across the cosmos. According to TOE, this isn't just an illusion of reference frames but a clue that resonance, not just geometry, underpins temporal coherence. The shifting present becomes a side-effect of gravitational waves influencing how time unfolds.

Time, in this model, is not fundamental. It emerges from the transformation of matter under gravitational resonance. Special Relativity describes its behaviour. TOE aims to explain its cause.

2. General Relativity

Einstein reframed gravity as the bending of spacetime by mass. TOE reframes that bending as resonance. The curvature isn't dismissed; it is recast as the effect of an inward-pulling gravitational wave interacting with the quantum field. Objects move not along bent geodesics but within a resonant structure that subtly pulls them inward.

This resonance doesn't merely shape orbits. It imprints coherence into the quantum world itself. Atoms, particles, and fields are shaped by the same gravitational rhythm that organises stars and galaxies.

3. The EPR Paradox and Entanglement

Einstein's discomfort with entanglement stemmed from its non-locality. TOE proposes that what links entangled particles is not mystery, but resonance. A shared oscillation across quantum space allows these particles to behave as one system. It isn't "spooky", it's structured. The gravitational field becomes the medium through which coherence is maintained.

4. Absolute Rotation and Frame Dragging

Experiments like Gravity Probe B showed that spinning objects drag spacetime. But what if this isn't space being twisted, but resonance being measured? The gyroscopes didn't just drift. They precessed against a deeper structure, an inward-pulling gravitational wave that provides a frame beyond relativity. Rotation becomes measurable not just in relation to other bodies, but against the quantum fabric itself.

TOE suggests that what we've interpreted as frame dragging may be evidence of this background field. The "warp" is a pull, not a twist. The mathematics still work, but the metaphor changes.

5. The Problem of Curvature as Metaphor

Curved space has become a foundational metaphor in physics. But metaphors, however useful, can obscure. Einstein used geometry as a placeholder, not an end. TOE replaces that placeholder with a physical mechanism: resonance. The geodesics remain. The equations hold. But they now describe rhythm rather than deformation.

Curvature describes local motion. Resonance defines the global coherence.

This shift matters. If gravity is resonance, then space is not an entity that bends. It is a structure that responds. The visual model of rubber sheets gives way to one of inward motion, an ongoing pulse that defines both matter and motion.

In Summary

Einstein gave us the courage to question the nature of time, space, and gravity. TOE builds on that courage. Curvature, in this view, is not a destination but a stepping stone. If mass shapes space, it is through resonance, not geometry. If time slows, it is due to the nature of transformation. And if entanglement is real, it is grounded not in paradox but in pattern.

We honour Einstein not by preserving his metaphors, but by pushing past them. What he saw as shadows on the wall, we now begin to trace back to the source. Not the warping of space, but the insight into gravity itself, guiding everything from the paths of planets to the coherence of the cosmos.

25. Penrose's Theories

Roger Penrose, a figure of profound and steady insight, has long contributed to the deeper architecture of physics. His ideas on quantum mechanics, spacetime geometry, and consciousness align naturally with the foundations of the Theory of Everything (TOE). One of his most overlooked yet crucial propositions is this: We need to "gravitise" quantum mechanics rather than quantise gravity, which carries profound implications, perhaps more than mainstream physics is yet willing to admit.

1. Twistor Theory and Toroidal Geometries

Penrose's Twistor Theory offers not a revision of spacetime, but a complete reimagining. Instead of the familiar fabric of space and time, he proposes a realm structured by twistors—mathematical forms that blend space, time, and momentum into a unified field.

While this might seem abstract, it finds resonance within the TOE, where gravitational light waves are not background effects but sculptors of reality. If gravitational resonance underpins the quantum field itself, then twistor theory may be describing the very structure of that field.

The presence of toroidal forms in magnetic fields, plasma vortices, and even galaxies hints at this deeper order. These self-sustaining curves suggest that an inward gravitational pressure may be responsible for nature's repeated geometries. If so, twistor theory may be more than a mathematical curiosity. It may offer the precise framework needed to model resonance mathematically.

2. Spacetime Geometry and Quantum Phenomena

Penrose has long insisted that spacetime is not passive. It participates. It forms. It influences. This complements the TOE's view that gravitational waves are not just ripples, but active agents shaping the quantum world from beneath.

His attempts to unify quantum mechanics and gravity point to this same conclusion: they are not disconnected systems but facets of a single underlying process. If gravitational resonance sculpts both atomic behaviour and galactic structure, then we are looking at a universe not curved by mass alone, but rhythmically shaped by an unseen frequency.

Across scales, from subatomic particles to spiral galaxies, we observe similar shapes. This recurrence may be the echo of gravitational resonance. An invisible structure emerging wherever there is enough coherence for it to take form.

3. Consciousness and Quantum Mechanics

Penrose has never shied from confronting the enigma of consciousness. His Orch-OR theory, developed with Stuart Hameroff, proposes that consciousness arises from quantum processes within microtubules. It is a contentious claim, yet it persists precisely because of its audacity and coherence.

If quantum mechanics shapes reality at every level, why exclude the mind? The TOE proposes that consciousness may be a product of resonance; formed in the same quantum crucible that shapes matter and motion. Under this lens, Penrose's idea becomes a suggestion that intelligence itself is not an anomaly but a resonance, emergent in systems complex enough to sustain it.

4. Cosmic Cycles and Gravitational Resonance

Penrose's Conformal Cyclic Cosmology (CCC) posits that the universe undergoes an endless sequence of aeons, each ending with the emergence of the next. Instead of fading into a heat death, the universe resets, stripping away distinctions between end and beginning at extreme scales.

This dovetails with TOE's vision of gravitational resonance as a fundamental driver of cosmic structure. If resonance underlies both expansion and contraction, then the so-called Big Bounce may not be a collapse at all, but a rhythmic transition—an echo in a cosmic metronome.

Conclusion

In integrating Penrose's vision into the TOE, we glimpse a deeper coherence. Twistor geometry, quantum gravity, gravitational resonance, and consciousness are not separate inquiries but reflections of one unified process. Penrose offers the mathematical language; the TOE supplies the physical context. Together, they shape a model that extends from the birth of galaxies to the emergence of self-awareness.

This synthesis reframes the story of the universe. Not as a machine governed by disjointed forces, but as a field of resonances: subtle, recursive, and alive with structure. In that sense, Penrose's greatest contribution may not be his equations, but his insistence that reality is stranger and more ordered than we have yet dared to realise.

26. Pauli Exclusion Principle

The Pauli Exclusion Principle tells us something deceptively simple: no two fermions can occupy the same quantum state. Electrons and protons are kept apart by a rule buried deep in quantum mechanics. And it matters. Without it, atoms collapse. Matter dissolves. But perhaps this isn't just a quantum formality. Perhaps it's not abstract at all.

The Theory of Everything suggests the principle emerges from the structure of spacetime itself, driven by a gravitational resonant wave pressing inward from a fourth spatial dimension.

Quantum Space as an Emergent Resonant Field

In this view, space isn't a blank canvas. It's textured, patterned, stirred into form by resonance. This resonance is not metaphorical. It arises from gravity, not as we commonly conceive it, but as a four-dimensional light-like wave constantly shaping the geometry of reality.

If that wave determines how matter and energy settle in three-dimensional space, then principles like Pauli's may not be constraints imposed from above. They're outcomes. Natural consequences of how fermions interact with the underlying spatial architecture. Overlap isn't prohibited. It's structurally impossible.

Resonance and Quantum States

Fermions have spin, but within this framework, spin may reflect deeper harmonic relationships. Each particle's quantum state aligns with a unique resonant frequency. These frequencies act as spatially embedded "slots," defined by gravitational resonance. The exclusion principle becomes a geometric necessity rather than a mysterious law. Identical fermions cannot share a state because their patterns cannot occupy the same resonance.

Gravity as the Architect of Quantum Restrictions

If gravity, as a 4D wave, sculpts the very layout of quantum space, then the Pauli Exclusion Principle is not a bolt-on rule. It's an imprint. A visible expression of an invisible framework. This reshapes the idea of quantum rules as arbitrary postulates. Instead, they are born from the deeper pressures of resonance acting upon the quantum field.

Matter and Energy: A Gravitational Blueprint

In this model, gravity is more than a large-scale force. It is a universal organiser. It sculpts matter and energy at every scale, from subatomic structure to the skeletal framework of galaxies. Fermions are not merely particles. They are harmonics within a resonant field. The exclusion principle is one way this structure preserves form and coherence, much as gravity preserves orbit and alignment on a cosmic scale.

Spin, Symmetry, and Spatial Harmony

Spin becomes more than a quantum descriptor. It reflects a particle's fit within a layered harmonic structure. Each fermion resonates with a particular phase of the gravitational wave. Their inability to share a state emerges not from command but from conflict: two identical resonance patterns cannot occupy the same spatial node. The exclusion principle becomes an expression of symmetry, not constraint.

A Unified Resonance

Within the TOE, the Pauli Exclusion Principle is no longer a mysterious rule imposed by quantum mechanics. It emerges naturally from gravitational resonance shaping quantum space. The same force that forms galaxies also structures atomic stability. This unification bridges scales. The exclusion of two electrons from a single orbital is governed by the same invisible wave that curves spacetime and maps the structure of the cosmos.

27. The Planck Constant: A Fundamental Harmonic of Reality

The Planck constant, h, is often treated as a simple scaling factor. A placeholder that links energy and frequency through the equation:

$$E = h\nu$$

It looks straightforward. But this relation reveals something profound: energy is not continuous. It arrives in steps. Quanta. This alone tells us something about the universe. It vibrates; not fluidly, but rhythmically, in discrete patterns. Which raises a deeper question: what is h really telling us?

In the Theory of Everything, h begins to look less like an input and more like an output. A result of something deeper. If energy is linked to frequency, then quantisation may arise from a gravitational resonant wave shaping the geometry of space.

Resonance as the Architect of Matter and Energy

Within this framework, resonance is not incidental. It is structural. Gravitational resonant light waves form the scaffolding from which both matter and energy emerge. The Planck constant, long seen as a passive coefficient, may instead reflect a constant of resonance, governing how energy is shaped and propagated within quantum space.

1. Energy Quantisation and Cosmic Resonance

That h connects energy to frequency implies that quantisation itself may be a resonant effect. Under the TOE, photons and particles emerge with energies defined by the oscillatory character of gravitational resonance. The Planck constant would then act as a bridge between microphysics and cosmic structure.

2. Photon Dynamics and the Expansion of Space

Standard theory holds that a photon's energy is defined by its frequency. But in this extended view, photons may also interact with gravitational resonance embedded in quantum space. If such waves influence how energy unfolds, they could also shape large-scale cosmic expansion. The Planck constant may encode not just local interactions but an imprint of a universal rhythm.

Resonance, Fluctuations, and Cosmic Structure

While h doesn't explicitly appear in the equations of cosmic expansion, the TOE reframes this absence. If gravitational resonance influences the fabric of space, then those early quantum fluctuations--seeded in the primordial universe--may have been guided by these deeper harmonics.

The constant may have acted as a tuning dial, amplifying particular frequencies. It would not have powered expansion directly, but it may have determined how the early structure of the universe formed, how fluctuations cohered into stars, galaxies, and the great cosmic web.

Primordial Fluctuations and Gravitational Structure

In the earliest moments, quantum fluctuations weren't peripheral. They were the universe. What shaped them? If they vibrated at specific frequencies, then h-as the key to energy-frequency relationships, may have played a hidden but central role. Not as a passive coefficient, but as a selector. A filter through which only certain patterns gained coherence and grew.

Gravitational Frequencies and Expansion

Within this framework, gravitational waves are not noise. They are resonant, structured, and spectral. If spacetime itself vibrates at specific frequencies, then the expansion of the universe may reflect this rhythm. The Planck constant could be more than a quantum bookmark. It may function as a regulator of cosmic dynamics, scaling quantum energy while also synchronising with the tempo of the cosmos.

Quantum Gravity Without Gravitons

If the TOE holds, there may be no need to quantise gravity using particles like gravitons. Gravitational resonant light waves offer a coherent alternative, linking gravity and quantum mechanics without contradiction. Here, h becomes a signature of this resonance. A tuning parameter for how matter and energy crystallise from a deeper order.

Energy, Frequency, and Resonant Structure

Gravitational resonance may represent a higher-dimensional organising principle. A wave-like architecture from which both energy and matter derive form. The Planck constant, with its precise relationship to frequency, would be the imprint of this structure. Not arbitrary, but inevitable.

Toward a Resonant Quantum Gravity

Attempts to merge quantum mechanics with gravity often force two mismatched systems together. But in this model, they are already entangled. If gravity operates as a resonant wave, then h underlies both systems, not as a compromise, but as a unifier. A constant that belongs equally to the quantum and the g.

Conclusion

The Planck constant does not cause the universe to expand. But it may guide it. Not as a force, but as a rhythm. A resonance. If this holds, h is not merely a footnote in quantum mechanics. It is a cipher. A key to a harmonic structure shaping every photon, every atom, every filament of space. The constant we once took as a scaling factor may turn out to be the tempo of reality itself.

28. Quantum Mechanics

Quantum mechanics, for all its predictive precision, remains conceptually unfinished. It excels at describing microscopic phenomena, yet struggles to integrate with gravity or classical physics. It relies on external assumptions: observers, postulates, and measurements-without resolving the mechanisms behind them. The Theory of Everything (TOE) offers a unifying thread: fourth-dimensional gravitational resonance.

The Measurement Problem: Collapse or Coherence?

At the centre of quantum theory lies the measurement problem. A system hovers in superposition until observed, at which point it 'collapses' into a definite state. But why? The standard model provides no internal cause. TOE reframes this: collapse is not enforced by observation but by resonance.

Detectors are not abstractions. They are matter, with mass, governed by gravity. If gravitational resonance extends into a fourth spatial dimension, then detection becomes an interaction within that resonant field. Measurement, in this sense, is not a metaphysical event but a physical alignment; a coherence imposed by gravity.

The Classical Lens: An Obstructed View

Despite its radical implications, quantum mechanics remains anchored in classical assumptions: space, time, and macroscopic detectors. It describes probability, yet offers no causal account of why or how collapse occurs. TOE proposes that gravitational resonance governs this transition. Detectors do not merely observe; they participate. Their structure resonates with quantum systems, guiding outcomes toward coherence. What we call 'measurement' is a synchronisation with the gravitational field, not an external imposition.

Gravity: The Missing Ingredient

Quantum theory, in its current form, excludes gravity. TOE places it at the heart. Superposition, entanglement, and wave-particle duality may not be anomalies but signs of gravitational tuning. If gravity shapes not only spacetime but the internal dynamics of quantum systems, then quantum behaviour arises not from uncertainty, but from structured resonance.

Entangled particles, then, are not linked by spooky action but by shared participation in a deeper gravitational rhythm. Collapse is not mysterious; it is a resonant stabilisation, the system settling into the only pattern permitted by the geometry of space itself.

Completing the Puzzle

Quantum mechanics remains empirically powerful, yet incomplete. Its dependence on unexplained assumptions suggests it is an effective model, not a final one. TOE offers a path beyond this impasse. If gravitational resonance structures spacetime, then quantum outcomes follow necessarily. There is no need for an observer to collapse a wavefunction; the structure of the universe does the work.

In this light, detectors and observers are not special. They are resonant bodies embedded within a field that governs all interaction. Collapse is not forced from outside, but drawn from within; gravity revealing its influence at the smallest scales.

The search for unity will not be won by patching quantum theory to general relativity. It will require recognising that both are approximations of a deeper process, one governed by resonance, not rules. TOE does not discard quantum mechanics; it provides the missing context. And within that context, the quantum world begins to make sense.

29. Particle Mass Patterns in the Standard Model

The mass hierarchy of particles in the Standard Model, where each generation scales upward in mass, suggests more than coincidence. The Theory of Everything (TOE) reframes this pattern as an emergent property of gravitational resonance shaping the quantum field. Mass, in this view, is not fundamental. It is structured.

1. Particles as Resonant Modes TOE proposes that particles are not discrete objects but stable standing waves in a four-dimensional quantum fabric. Quarks, neutrinos, and other so-called particles emerge as oscillatory modes shaped by inward gravitational light waves. Mass arises from the energy density of these resonant fields. What we perceive as matter is the visible edge of a deeper spatial harmony.

2. Confinement, Imaging, and Interference Quark confinement becomes less mysterious if quarks are held in place by resonance fields rather than by raw force. TOE also suggests that many subatomic events are imaging effects: interference patterns in quantum space misinterpreted as particle collisions. Neutrino oscillations, for instance, may reflect resonance shifts, not intrinsic changes in mass.

3. Compression, Mass, and Quantum Space Mass can be seen as a product of gravitational compression within quantum space. This compression increases energy density and gives rise to the illusion of solidity. Rather than being passive, gravity shapes matter directly. Einstein's curvature and quantum fluctuations intersect here, unified by resonance.

4. Neutrinos and Exotic Resonances In this model, neutrinos are not particles in the traditional sense. They are resonant signatures; waveforms shaped by gravitational fields. Passing through crystalline media like ice may briefly stabilise these forms, making them detectable. But that doesn't mean they are travelling objects. They could be quantum echoes, brought momentarily into coherence by field alignment.

The 2024 detection of a 220 PeV neutrino with no clear astrophysical source raises this question. If not from decay or collision, what caused it? TOE suggests it may have been a gravitational disturbance, briefly stabilised into detectability when field resonance aligned with the detector's structure.

This parallels how cold temperatures preserve coherence in quantum systems. Just as chill reduces decoherence in qubits, ice or ultra-pure water may stabilise oscillations long enough for observation.

Similarly, exotic particles like tetraquarks or pentaquarks may be transient resonances rather than new forms of matter. If so, no additional forces or speculative particles are needed. Gravity's resonance explains their brief existence.

5. Detection, Assumptions, and the Neutrino Problem Neutrinos have never been directly observed. Their existence is inferred through missing energy, water light patterns, or secondary effects. Their mass is uncertain, and their behaviour defies classical categories. They switch identity mid-flight, more like harmonics than particles.

TOE treats these anomalies as signs of resonance. Nuclear events and quantum transitions may create wave interference that appears briefly particle-like. What we detect are stabilised field disturbances, not persistent objects.

6. Beyond the Particle Paradigm

TOE challenges the foundations of particle physics by redefining what a 'particle' is. These are not indivisible units of matter, but emergent field patterns. There is no need to postulate particles like gravitons. Gravitational resonance accounts for structure, stability, and even anomalies often attributed to dark matter, which may result from spatial compression or optical misinterpretation.

This approach unifies high-energy physics, quantum field theory, and gravity under a single principle: resonance. It dissolves artificial categories and reframes much of particle physics as a projection of deeper interactions.

Summary Particles, under TOE, are not fixed units of substance. They are shaped by a deeper field. Gravity, expressed through higher-dimensional resonance, does not merely structure the cosmos. It shapes the atom. The complexity of particle physics is reinterpreted as emergent geometry, a more straightforward and more coherent concept.

This dissolves the illusion of fundamental objects, revealing a resonant field of energy and form. What we've taken for substance may be fleeting alignments, moments of coherence in a vibrating medium. TOE doesn't multiply entities or invent exotic particles. It simplifies. It listens to the structure of space, not just its content. In doing so, it invites a new way of seeing matter.

30. Triplication of Nature's Fundamental Constituents

Nature repeats itself in threes: three generations of quarks, three generations of leptons. Nearly identical, differing only in mass. It's a structure physics has long described but never truly explained.

From the TOE perspective, this triplication may not be arbitrary. It may reflect resonance. These particle families could represent harmonics. Three stable states anchored within the quantum-gravitational field. Just as planetary orbits fall into stable configurations under gravity, these particle generations may correspond to discrete resonant nodes embedded in the geometry of spacetime itself.

This threefold layering need not be mysterious. It mirrors a deeper order seen across scales: from atoms to galaxies, nature seems to favour balance and patterned coherence. Each generation may occupy a specific energetic tier, stabilised by resonance, like chords in a deeper spatial harmony.

Rather than a glitch or coincidence, this structure could reflect a quantum harmonic ordering, where gravitational resonance naturally gives rise to three distinct yet interconnected tiers. The universe, it seems, prefers elegant repetition. Not forced, but formed. The triplication is not a mathematical afterthought. It is the outcome of structure seeking coherence.

31. Quantum Entanglement

Quantum entanglement has long defied explanation. Particles, separated by vast distances, appear to influence each other instantaneously. It seems to violate locality, suggesting a kind of communication that outruns light itself.

But perhaps it's not communication at all. From the TOE view, entangled particles are not truly separate. They remain bound through a deeper geometry; one that stretches beyond our three-dimensional frame. The illusion of distance arises because we're observing from within a limited dimensional slice of a more complex structure.

Imagine a two-dimensional world trying to understand a three-dimensional object. It would see separate projections of what is, in truth, one form. So too with entanglement. The connectedness is not spooky, it's structural.

TOE frames this as coherence within a 4D gravitational resonance field. The apparent nonlocality is not a violation of physics but a signpost. Entangled systems remain part of a unified wave structure, their correlation a natural consequence of shared resonance across higher-dimensional space.

Entanglement, then, may not be an anomaly. It is a clue, a reflection of a universe whose fabric is profoundly and fundamentally connected.

32. The Measurement Problem and Quantum Resonant Collapse

The Measurement Problem in quantum mechanics has often been framed as a paradox: a wave function collapses into a definite state only when observed. But perhaps this is less a fundamental riddle and more a misreading, rooted in classical assumptions and limitations.

1. Measurement Tools and Gravitational Resonance

Classical instruments do not measure absolutes. Atomic clocks, for instance, do not capture time itself. They register oscillations—fluctuations that may be shaped by underlying gravitational resonance. Like a barometer reacting to air pressure, these devices respond to shifting gravitational conditions. The apparent randomness of quantum outcomes might, in part, reflect these fluctuations in the background field, not an intrinsic indeterminacy.

2. Measurement as Resonant Alignment

TOE proposes that what we call "collapse" is not a sudden or arbitrary shift, but a resonance event. A quantum system, embedded in a field of gravitational oscillations, aligns with stable resonant frequencies when measured. Only those states that harmonise with the surrounding gravitational structure are amplified into observability.

3. The Observer as a Resonant Participant

In this view, measurement does not require a conscious observer imposing change. The act of observation becomes a passive resonance alignment between detector and system. Measurement is not an intrusion, but a convergence where quantum probabilities resolve through coherent interaction with the gravitational field.

4. Collapse Reimagined

The wave function need not collapse in the dramatic sense. It may simply settle. Out of many potential states, one emerges, not because it was selected by observation, but because it fit most cleanly into the local gravitational resonance. Stability, not intervention, defines the outcome.

5. Rethinking Quantum Reality

This reframes quantum behaviour not as a realm suspended in ambiguity, waiting to be collapsed by consciousness, but as a field already biased toward certain outcomes. Observation does not cause a result. It reveals the one already most coherent with the resonant geometry of space.

6. Decoherence and Resonance Breakdown

Decoherence, often attributed to environmental noise, may instead signal a breakdown in gravitational resonance. As a quantum system interacts with its surroundings, coherent harmonics begin to fragment. What follows is not disorder, but a transition: an adjustment into the stable, familiar patterns of classical physics.

Conclusion

The TOE reframes the measurement problem as a feature of resonance rather than a mystery of consciousness. What appears as collapse is the universe resolving itself through structure, not surprise. Measurement is no longer a metaphysical anomaly. It is a natural outcome of a universe, seeking equilibrium through gravitational order.

33. Quantum Scarring: Patterns in Chaos

Quantum scarring challenges the idea that quantum systems are wholly random, especially in chaotic regimes. In theory, wavefunctions in such systems should spread evenly, with no memory of classical behaviour. And yet, scars remain; concentrated probability regions that trace classical paths long thought irrelevant.

These scars show up as subtle, structured remnants. Instead of dispersing uniformly, quantum systems retain a ghostlike alignment with the trajectories of classical motion.

Experiments with graphene quantum dots have directly revealed this. In chaotic geometries, wavefunctions persist along classical paths, suggesting an underlying pattern. So what causes this? Are we just seeing artefacts of boundary conditions? Or is something deeper enforcing this order?

Resonance as Structural Guidance

In TOE, resonance is not a side effect. It shapes everything. Gravitational resonance, subtle and continuous, may be influencing the structure of quantum systems at a foundational level. Scarring, from this view, is not an error or an exception, but a signature of this deeper organising force.

Rather than randomness, what we are witnessing may be alignment. Quantum scars become visible markers of gravitational resonance, shaping the flow of energy and probability. What looks like noise may be patterned. Structured. Resonant paths made visible in a system that still appears chaotic.

Biological Resonance

If gravitational resonance can stabilise quantum wavefunctions, it may also play a role in more complex systems. Protein folding, for example, remains poorly understood despite computational advances. Nature achieves it in milliseconds, not through calculation but through something more inherent.

Within the cell, where energy, geometry, and chemistry intersect, even small resonant influences could guide molecular structure. If this is correct, then biology too may reflect deeper gravitational harmonics. Life organising itself through resonance, not randomness.

A Reconsideration of Chaos

Quantum scarring may not be a curiosity. It may be a clue. A crack in the surface of standard theory through which a deeper coherence becomes visible.

In TOE, what we call chaos may often be resonance not yet recognised. These scars, lingering where none should remain, suggest that systems: quantum or biological, are not drifting aimlessly, but are shaped by an underlying field. A structure that reaches beyond the mathematical surface, quietly tuning the universe into form.

34. Beyond the Speed of Light

If the TOE holds, some quantum events, entanglement in particular, may not be unfolding entirely within the spacetime we recognise. They may be slipping through a deeper structure, one where distance loses its conventional meaning. Not by exceeding the speed of light, but by moving through dimensions we don't directly perceive.

Entangled particles might use geometric shortcuts, higher-dimensional resonant paths invisible to our classical framework. It's not magic, it's a misreading of the architecture.

And if these fields underlie all matter, the next question becomes sharper: could something conscious, something intelligent, learn to align with them? Biology may already flirt with this possibility. Migratory birds, through magnetoreception, could be tapping into quantum entanglement. Life, through evolution, might have stumbled upon these deeper structures long before we understood them.

But intelligence may take it further. Not just adapting, but tuning deliberately. Matching frequency with field. At some threshold, life might cease to be a passive passenger in spacetime and begin to move through it in ways matter alone cannot.

Gravity still shapes the universe at large, but quantum systems hint at a deeper set of rules. If these resonant pathways exist, nature may already be using them. The TOE does not ask us to discard the speed of light, only to look beyond the surface of space itself.

35. Atom

The classical image of the atom as a miniature solar system is elegant, but incomplete. Quantum mechanics shows us a stranger truth:

electrons don't orbit, they exist as wavefunctions, probability clouds anchored within an invisible lattice of forces.

The forces at play are not merely outward or inertial. They are inward, converging. Electromagnetism binds electrons to nuclei. The strong force holds protons and neutrons together. And gravity, reimagined through the TOE as an inwardly propagating resonant wave, may underlie them all.

This inward resonance provides an alternative way of understanding atomic stability. Instead of imagining electrons as tiny planets held in place by momentum, think of them as waveforms suspended in a field shaped by gravitational coherence. Not static, but in dynamic equilibrium.

Two implications follow:

1. The electron's position is not fixed, but it is not random either. It exists within a sculpted resonance, shaped by both electromagnetism and deeper gravitational fields.

2. The atomic nucleus, a dense cluster of positively charged particles, should, in theory, explode apart- yet it doesn't. TOE suggests gravitational resonance at the nuclear level may offer a compressive counterforce, anchoring the nucleus against electrostatic repulsion.

In this light, the atom becomes more than a fundamental unit of matter. It becomes a window. A microcosm. A resonant structure where quantum, gravitational, and electromagnetic forces interlace. If the same inward principle echoes across the scales, from quarks to galaxies, then the atom might not just be made of the universe. It might mirror its deepest form.

36. Conservation of Momentum (TOE Interpretation)

Conservation of momentum is one of those principles we rarely question. Objects in motion stay in motion unless acted upon. Simple. Reliable. But perhaps deceptively so.

From the TOE perspective, this isn't just a mathematical convenience. It may reflect a deeper symmetry within the resonant architecture of space. Gravity, reimagined as an inward, four-dimensional wave, does not pull. It shapes. It creates rhythm; a continuous inward flow that guides matter rather than dragging it.

Momentum, then, might not belong to the object at all. It could be a resonance effect. A particle or body stays in motion not because of inertia, but because it remains attuned to this gravitational rhythm. Motion is not conserved: it is maintained through alignment.

Even in extreme environments such as black holes or neutron star collisions, momentum holds. Perhaps not due to an unbreakable rule, but because the background resonance never ceases. The dance continues, and objects move in accordance with it.

At the quantum scale, this takes on stranger forms. Particles tunnel, superpose, and flicker into view. They seem to defy classical motion. But if we view them through the lens of gravitational resonance, this may be what motion looks like at a scale where resonance governs and classical mechanics loses footing.

Momentum becomes less a law and more a footprint of a deeper, continuous pattern. Not static conservation, but fluid coherence, carried by the unbroken hum of space itself.

37. Gravity as a Quantum Field's Source in the TOE Framework

Physics often treats gravity as the outsider, a classical holdover waiting to be quantised. TOE reverses this view. Gravity, in this framework, is not subordinate to quantum fields. It is their source.

This inward, resonating wave does not require quantisation. It is already the sculptor. Quantum space emerges from its rhythm. Fields, particles, and probabilities take form within its curvature. Gravity, in this view, is not a force to be reconciled with quantum mechanics. It is the fabric through which quantum phenomena arise.

Quantum effects such as entanglement, uncertainty, and field fluctuations may be nothing more than expressions of this underlying gravitational resonance. What appears chaotic or probabilistic could be the result of observing a deeper order through a narrow lens. The vacuum itself, rather than being a froth of virtual particles, might be a structured harmonic medium.

If this holds, the decades-long pursuit of quantising gravity may be a detour. Not because gravity resists quantisation, but because it precedes it. Quantum behaviour may be downstream of gravitational resonance, not its foundation.

This turns the standard model on its head. Gravity is not the weakest force. It is the most fundamental, subtle, pervasive, and woven into the very act of existence.

38. Grand Unified Theory (GUT) as an Emergent Property within the TOE Framework

For over a century, physics has sought unification by stitching the strong, weak, and electromagnetic forces into a single framework. Gravity is usually added later, as an afterthought. But this approach may start from the wrong end.

In the TOE model, unification isn't something to be constructed. It unfolds. It emerges from something more primary: gravity. Not gravity as a force among others, but as the resonant medium.

This is not textbook gravity. It is inward, dimensional, a resonant wave shaping quantum space from the inside out. Not one force among many, but the condition that gives rise to the others.

From this view, the so-called fundamental forces are not separate phenomena to be reconciled. They are different expressions of the same underlying geometry, sculpted by gravitational resonance.

Resonance, not reductionism

The electromagnetic force becomes a ripple on a gravitational sea. Its charge interactions and photon exchanges follow patterns set by the rhythm of space, shaped by gravity's inward pull.

The weak force, with its asymmetries and short-lived interactions, may reflect gravitational gradients: a kind of field chirality, where slight shifts in resonance push particles toward decay or transformation.

The strong force, the most intense, appears where resonance compresses quantum space into tight, coherent loops. This is not simply confinement. It is geometry under pressure, the field folding in on itself.

Rather than demanding a unification through high-energy equations, these forces reveal themselves as different moods of the same field. Emergent, not engineered.

Symmetry, and why it breaks

In standard models, symmetry breaking is invoked whenever patterns fail to hold. But it often lacks an explanation beyond mathematical necessity.

In TOE, symmetry is a result of resonance. It arises like a pattern on a vibrating surface. When that resonance shifts, symmetry changes. Not abruptly, but fluidly. Not because something breaks, but because the wave bends and the field follows.

This is not a clean fracture. It is an evolution. A tuning. Symmetry, in this framework, is dynamic, sensitive to gravity's subtle rhythm.

Quantum fields as echoes of gravity

Quantum fields do not exist in isolation. In TOE, they are shaped and conditioned by the gravitational medium they inhabit. They are more like resonant chambers, formed by inward waves sculpting the quantum structure.

Each field carries its own harmonic signature. Particles are the standing waves, coherent configurations that survive long enough to be observed. Forces emerge through the ways in which space responds when gravity moves through it.

Why traditional GUTs fall short

Conventional GUTs rely on immense energy scales, far beyond our reach. They assume that, at some threshold, the forces will merge.

TOE suggests they already have, not at high energies, but at the level of resonance. The unification is not in the collision. It is in the pattern.

If gravity structures the field itself, then what we call forces are simply different harmonics of the same resonance. What appears as diversity is really variation. What looks like separation is only a limitation of our perspective.

The flaw in GUTs is not that they failed to unify, but that they assumed the forces were ever truly separate.

39. String theory, LQG, and the middle path

Now, what about the big contenders, String Theory and Loop Quantum Gravity?

TOE doesn't reject them outright. It absorbs what they got right.

LQG talks about loops, discrete, self-organising structures in quantum space. TOE reframes them. Those loops? They're resonances. Not quantised by default, but shaped by gravitational rhythm. Discrete, because gravity creates coherence in an otherwise chaotic field.

And strings? TOE sees them too. Not as fundamental units, but as emergent shapes. Vibrations on the surface of something deeper. Like standing waves in a drum skin. They exist because gravity sculpts the surface, not the other way around.

Crucially, in TOE, strings can't reach into the fourth dimension. That's reserved for gravity. Which means string theory, in its traditional form, only ever glimpses the shadow of the real engine.

So, where does that leave us?

1. Gravity isn't just another force; it's the origin point. The pulse behind the curtain.

2. All other forces? Just different ways space reacts when gravity moves through it.

3. Symmetry, and its breaking, is just a natural consequence of gravitational resonance shifting under stress.

4. Unification doesn't require exotic energies or twelve-dimensional math. It's already here, if you know where to look.

5. And yes, TOE might finally offer a bridge, not just between GUT and gravity, but between the best intuitions of LQG and String Theory.

We stop forcing unification. We start listening. Watching how the quantum field bends and breathes under the influence of something much older, much deeper. Something that doesn't need to be quantised because... it already is everything else.

40. The Great Event: Big Bounce, Big Bang in the TOE Framework

We tend to imagine the Big Bang as a singular, all-or-nothing event. A cosmic "*once upon a time.*" But what if that's not quite right? What if the universe doesn't begin or end, but breathes?

In the TOE framework, there is no true beginning. No end, either. Just a great pulse between expansion and contraction. A kind of cosmic heartbeat, written in black holes, gravitational resonance, and quantum waves. Not a firework followed by silence, but a rhythm; inhale, exhale, repeat.

The "Great Event" is the pivot. The switch. The moment when the universe folds in on itself, or bursts outward again. Not an explosion, but a retuning. A restart, not a start.

Galaxies and the Role of Black Holes If you follow the thread of this idea, galaxies dissolve into supermassive black holes, not in violent collapse, but through slow gravitational digestion. These black holes are not endpoints, but transitions. Once their quantum emissions fade, gravitational resonance takes over. Everything begins to shift.

This is where the flatness problem dissolves. You don't need exotic inflation fields or finely tuned conditions. Just gravitational resonance, softly realigning quantum space. The near-flat geometry we observe is not coincidence. It's maintenance.

And black holes don't simply vanish through Hawking radiation. In this model, they initiate contraction. They apply pressure. When their quantum hum quiets, the universe turns inward.

The Reversal: Expansion to Contraction Eventually, expansion slows, not due to any external resistance, but because internal energy-quantum elasticity begins to thin. Gravity holds its grip. Resonance intensifies. Contraction begins.

This reframes the horizon problem. In a cyclical model, resonance does more than shape structure; it carries information forward. It imprints memory into the next universe. Even regions that appear disconnected in this cycle retain striking similarities. They share a history. A kind of cosmic ancestry.

The Gravitational Resonance Event

When black holes reach critical density, space-time begins to curve inward. Tension rises. Then resonance peaks.

This is the reset. The Big Bounce. Or perhaps the next Big Bang. Not just a return of matter, but a reinflation of quantum space. The geometry resets. The rhythm begins again.

Redshift and Resonance

We usually interpret redshift as a sign of distance. But in TOE, redshift is linked to gravitational resonance. As resonance weakens across vast distances, light stretches. Wavelengths elongate. And as the universe contracts, resonance strengthens. A reversal occurs. A cosmic blueshift; one we might detect.

Redshift becomes more than a marker of expansion. It is a clue in the universe's breath cycle.

The Inflation Conundrum

Inflation theory has always felt a bit too convenient. TOE sidesteps it.

There's no need for artificial bursts. In a cyclical model, conditions reset naturally. Matter and energy redistribute. Uniformity emerges through rhythm, not speed. The universe, over time, learns to balance itself. A neater solution, by being slightly messier.

Beyond Heat Death: A Regenerative Cosmos

What if the universe doesn't end in silence? Heat death: the slow fading of all energy is a bleak vision. But TOE offers something else.

In this framework, contraction generates energy. Collapse becomes renewal. Each pause sets the stage for a crescendo. There is no final equilibrium. Only a breath, held, then released again.

Reinterpreting Eternal Expansion Eternal expansion has a poetic appeal. But TOE proposes something richer. A universe that expands, then contracts. A balanced rhythm. Not driven by dark energy, but by the interplay of gravitational resonance and fading quantum pressure.

The Multiverse Possibility

What if each "Great Event" doesn't just reset this universe, but births a new one? A different cosmos, born from new gravitational conditions.

TOE suggests a multiverse, not of clones, but of variations. Each cycle is a new improvisation. We may be one chapter in a much larger story.

The Great Attractor and Mass Reversion

Consider the Great Attractor, pulling galaxies toward a mysterious destination. TOE sees this as a symptom of deep gravitational resonance. Perhaps the largest black holes, upon reaching critical mass, flip the universe's polarity. Their pull overrides expansion. Everything turns inward.

Time itself begins to reverse.

Black holes, in this view, are not graves. They are initiators. Quiet engines that stir the next beginning.

Black Hole Formation in TOE

And singularities? TOE denies their infinite nature. Black holes are dense, resonant pockets. Fossils from previous universes. Seeds. These may survive the Great Event, carrying a faint blueprint into the next cycle.

Galaxies don't emerge from chaos. They inherit structure. Spirals, filaments, patterns: echoes from before. The next universe doesn't begin from scratch. It rebuilds from ruins.

Conclusion: A Regenerative Universe

So, what remains?

A cosmos that doesn't burn out. One that doesn't need metaphysical miracles. Just gravity. Resonance. Cycles.

TOE reframes the story. It doesn't patch redshift, flatness, or inflation. It recasts them. The universe becomes a self-renewing system: expanding, contracting, breathing, beginning again. And perhaps, some faint trace of those earlier lives still lingers. In the radiation. In the black holes. In the silence between stars.

Not a single beginning. But a long memory, still unfolding.

Chapter 15
The Theory of Everything IV

41. The Hierarchy Problem and Gravitational Resonant Light

There's something awkward about the Higgs boson. Not the particle itself, but what it implies. By all accounts, its mass should be enormous. Quantum corrections, left unchecked, should drive it toward the Planck scale. Yet it remains curiously light. This mismatch is known as the Hierarchy Problem: why doesn't the Higgs collapse under its own quantum fluctuations?

Standard physics suggests solutions through fine-tuning or supersymmetry. The TOE offers something different. It indicates the answer is not hidden in new particles or adjustments, but in gravity itself, as resonance.

Inward Resonant Gravity as Structural Light

In the TOE, gravity is not merely curvature. It is a resonant wave, an inward-moving structural intensity; a kind of light, not made of photons, but of geometry under compression. As mass increases, this inward resonance amplifies. It is not just acting on matter; it shapes it. Every particle is immersed in this resonant field, every interaction conditioned by its tone.

This feedback loop. Mass-shaping resonance, resonance-shaping field. Creates stability not through force, but through coherence.

The Higgs as a Product of Resonance

This reframes the problem entirely. The Higgs mass appears light not because of cancellation or symmetry protection, but because it is stabilised by resonance. The gravitational field, permeating the quantum vacuum, acts as a dampener on wild fluctuations. It restrains excess, narrows variation, and holds the Higgs in place through geometric balance.

There's no need for counterweights. The resonance itself prevents runaway behaviour. The Higgs is light because the universe tunes it that way.

A Bridge Between Fields and Gravity

This reshapes the deeper question. If gravity is a resonant medium that informs quantum fields, then perhaps the fields aren't separate at all. Mass may not simply react to gravity; it may arise from it. The Higgs, the electron, the field structure of matter itself, could all be resonant expressions within this inward gravitational wave.

In that light, the attempt to quantise gravity may be misplaced. What if quantum behaviour is already gravitationally defined? What if resonance precedes the field?

Then gravity is no longer an outlier to be reconciled. It is the frame. The Higgs, far from being a mystery, becomes evidence of a universe where form, mass, and balance all emerge from a deeper gravitational phenomenon.

42. TOE's Broader Implications for Unsolved Problems in Mathematics

Physics may be the canvas, but mathematics has long claimed to be the language behind it. Yet if the Theory of Everything reveals something deeper, something structural, resonant, embedded in the architecture of space itself, then perhaps mathematics is not merely descriptive. Perhaps it is emergent.

At the core of TOE is the idea that space, time, and quantum phenomena arise from an inward gravitational resonance. A vibrational geometry that does not just describe the world, but generates it. From this view, mathematics may not exist apart from the universe, but within it: an echo of physical law resonating outward into the conceptual.

Unsolved mathematical problems, such as the Riemann Hypothesis, prime distribution, or P vs NP, may not remain unresolved due to intrinsic difficulty, but because they are being asked within the wrong dimensional framework. These puzzles might be shadows cast by higher-dimensional resonances that classical logic cannot fully trace.

If resonance underpins quantum coherence, and space emerges from curved, inward-propagating gravitational light, then number theory may reflect this structure. Prime numbers, often treated as chaotic, might align with interference patterns across the quantum vacuum. What appears random in flat mathematical space can be ordered when viewed from the perspective of resonance curvature.

This does not undermine mathematics. It grounds it. Rather than a sovereign abstract domain, mathematics becomes a resonance map, conceptual tones shaped by the structure of space itself. The apparent limitations of mathematics may not be epistemological, but geometric.

Take the Riemann Hypothesis. Its elusive zeroes may correspond to tonal nodes in resonant space, not simply artefacts of complex analysis. Or Goldbach's Conjecture: perhaps it reflects a deeper pairing symmetry woven into the fabric of number space. Even Gödel's incompleteness theorems begin to feel aligned with this idea: that no system can fully account for itself if the system itself is in motion. If resonance shapes thought as well as form, then incompleteness might be a natural feature of oscillating foundations.

TOE does not solve these problems. But it reshapes their contours. It suggests that many of our mathematical impasses are not failures of logic, but failures of perspective. What we may need is not tighter proofs, but a deeper understanding of the chamber from which these mathematical echoes emerge.

In this light, the next breakthroughs in mathematics may not stem from abstraction, but from alignment, with a geometry that computes beneath both number and nature.

A List of Validated Theories the Theory of Everything Complements or Extends Upon

The Theory of Everything (TOE) isn't interested in starting a fire. It's not a demolition of the house that physics built, but rather an attempt to walk through its halls with a deeper awareness of what lies beneath the floorboards. Much of what classical physics has given us remains intact, formally accurate, elegantly structured, but perhaps incomplete. TOE doesn't reject these foundations, it simply suggests there's more, something hidden in the resonance between what we observe and what we've yet to conceptualise.

1. Special Relativity (Einstein)

Time dilation and mass-energy equivalence hold, but TOE reframes time itself. Rather than a standalone dimension, time becomes a property emergent from higher-dimensional resonance: a rhythm shaped by gravitational interaction, not a static parameter.

2. General Relativity (Einstein)

Einstein's general relativity remains one of the most elegant and accurate descriptions of large-scale motion, planetary orbits, and black hole dynamics. However, within the TOE framework, the concept of spacetime curvature becomes a useful surface projection that conceals a deeper, more fundamental process. Rather than imagining gravity as a bending of space, TOE sees it as the result of inward, resonant waves shaping quantum space itself. These waves do not warp space in the visual metaphor of rubber sheets; they structure it from within, like tension shaping a drum skin. The geodesics of relativity still hold, but here, they follow not a curved void but the contours of resonant density. General relativity is not discarded; it is absorbed. Accurate within its scope, but emerging from a deeper order of reality.

3. Lorentz-Related Extensions in the TOE

TOE respects core Lorentz principles but extends them into quantum and higher-dimensional domains, especially under extreme conditions.

Lorentz Symmetry TOE maintains invariance across inertial frames in conventional settings but allows for small, context-specific deviations, especially near black holes or at quantum scales, arising from interactions with gravitational resonance.

Lorentz Invariance Violation (LIV)

Where violations appear at high energies or in quantum anomalies, TOE interprets them not as fundamental breaches, but as local effects of gravitational wave interference. Symmetry holds globally, even if it flexes locally.

Lorentz Force Classically tied to charged particle motion in electromagnetic fields, the Lorentz force may be subtly influenced by gravitational resonance in extreme regimes. This could refine our models of charge-field behaviour at the quantum boundary.

Lorentz Covariance The TOE retains covariance across Lorentz transformations, while recognising that gravitational resonance may influence quantum field dynamics without undermining the symmetry of physical law.

Higher-Dimensional Symmetry In higher-dimensional space, TOE extends Lorentz principles into a dynamic geometry, offering new paths toward explaining entanglement and rethinking the physics of regions currently thought to be singular.

4. Electromagnetism (Maxwell's Equations)

Maxwell's framework remains intact, but TOE reframes electromagnetism as a harmonic effect rather than a standalone field. Rather than collapsing it into gravity, TOE treats it as a resonance shaped by the same inward gravitational geometry. Electromagnetic waves become ripples modulated by a deeper, dimensional field; two expressions shaped by the same medium.

5. Thermodynamics

TOE keeps the laws but reinterprets their meaning. Entropy is not just statistical drift but a resonant shift. Energy moves not randomly, but toward equilibrium within a gravitationally structured field. Heat flow, dissipation, and the arrow of time may reflect how energy settles into resonance with quantum space itself.

6. Pauli Exclusion Principle

TOE sees the exclusion principle as arising not from restriction, but resonance. Fermions differentiate not by force, but by pattern, each held in a distinct phase by gravitational structure. Matter holds form not through repulsion, but through harmony enforced by the deeper geometry of space.

7. Quantum Mechanics (Schrödinger, Heisenberg)

Wave-particle duality and quantum uncertainty need not imply randomness. In TOE, they are resonant structures, not mysteries. What looks like chaos may be interference from deeper gravitational waves shaping subatomic geometry. Schrödinger's wavefunction becomes a harmonic map, not a probabilistic cloud. Quantum indeterminacy may simply reflect limited access to a higher-dimensional coherence.

8. Quark Model (Gell-Mann, Zweig)

Quarks are not indivisible particles, but vibrational nodes in a deeper resonant field. Their confinement may reflect gravitational resonance, not just strong force equations. Spin, colour, and charge emerge from the geometry of space under pressure. The model holds, but TOE gives it scaffolding, revealing quarks as stable formations in a field tuned by gravitational structure.

9. Bell's Theorem

Bell's violations need not abandon realism. TOE reframes them through dimensionality. Entanglement isn't strange; it's structural. Particles behave as if linked because they are, through a shared resonance. What appears to be nonlocal behaviour may be harmonics across a geometry that we cannot fully see.

10. Quantum Field Theory (QFT)

QFT treats particles as field excitations, but remains silent on what shapes the fields themselves. TOE proposes gravitational resonance as the sculptor, an inward compression from a higher dimension that gives structure to quantum fluctuations. The vacuum is not empty, but textured with this resonance, shaping chaos into form.

11. Higgs Field and Higgs Boson

The Higgs field explains how mass is assigned, not why it exists. TOE suggests mass arises not from a metaphysical mechanism, but from gravitational resonance. A four-dimensional inward wave shapes the Higgs field, tuning it like a drumhead. The boson is not a magical origin, but a resonant peak in a larger field. Mass may be geometry under pressure, not an intrinsic trait.

12. Evolutionary Biology (Darwin)

Darwinian evolution relies on randomness filtered by selection. TOE offers an added layer: resonance as a shaping influence. Biological complexity may reflect gravitational tuning at molecular and cellular scales, guiding the emergence of order from chaos. Not a rejection of Darwin, but a deepening of his insight.

13. Fine Structure Constant

This silent number sets the strength of electromagnetism, long treated as a mystery. In TOE, it may emerge from the resonant harmonics of space itself, a fingerprint of gravitational geometry embedded in field behaviour.

14. Geodesics (General Relativity)

Einstein's geodesics trace the curves of spacetime. TOE suggests those curves are not static but subtly adjusted by inward gravitational resonance. The paths of matter reflect not only mass, but coherence, shaped from within.

15. Penrose's Theories

Penrose's twistors, non-computable mind, and elegant geometries have often seemed esoteric. But under TOE, they find grounding. Consciousness becomes an echo of deeper resonance; geometry, perhaps, a living structure. What once felt fringe now reads as framework.

16. Big Bang Theory or Bounce

Rather than a singularity from nowhere, TOE points to a bounce: a compression followed by expansion, tuned by gravitational resonance. The universe may not have exploded into being; it may have harmonised into form.

17. Chaos Theory

TOE reframes chaos as patterned complexity. The apparent randomness of nature may be a layered resonance playing out through sensitive systems. Complexity is not a deviation but an expression of the underlying wave.

18. Schrödinger Equation

Wavefunctions don't evolve in a vacuum. They ripple across a medium. TOE suggests that this medium is not empty space but a resonant quantum lattice, gently shaped by gravitational light waves. Schrödinger's equation still holds, but now with context. The oscillating wave isn't an abstract probability cloud. It's the echo of matter conversing with the very geometry of space-time.

19. Heisenberg's Uncertainty Principle

Not a measurement flaw, but a reflection of interaction. In TOE, uncertainty arises from particles brushing against the gravitational field's deeper hum. The blurring of states is not an error but a feature of higher-dimensional resonance.

20. Standard Model Lagrangian

Accurate, yet incomplete. The Lagrangian captures interactions but not the deeper architecture. TOE suggests that gravitational resonance shapes these interactions from beneath, acting as the silent structure guiding the field's choreography.

21. Dirac Equation

More than a predictor of antimatter, Dirac's work hints at symmetry beneath complexity. TOE extends this by showing how gravitational resonance shapes mass and spin, embedding quantum structure in a coherent, resonant field.

22. Fourier Transform

More than a method, it is a decoder of resonance. TOE sees every interference pattern as a harmonic, each fluctuation as a note. The Fourier lens becomes a way to glimpse the structure behind appearance, if tuned to the right dimension.

23. Euler's Formula

Euler's formula reveals an elegant symmetry between exponential growth and rotation, as well as between the real and imaginary components. More than mathematical beauty, it hints at something physical. In three dimensions, it traces a helix, a form echoed across nature, from DNA to galaxies. This spiral may reflect a deeper principle, where oscillation and form co-emerge. Euler's identity, uniting five constants in one equation, seems less like a coincidence than a window into the universe's underlying coherence.

24. Kepler's Laws of Planetary Motion

Kepler's ellipses have long described orbital paths, but TOE suggests a deeper structure. Planetary motion may be more than gravitational mechanics: it may be the outcome of resonance patterns in space itself, with each orbit shaped not only by mass and momentum, but by harmonic fields woven through the quantum medium.

25. Virial Theorem

The balance between kinetic and potential energy describes dynamic systems at rest. TOE reframes this as the visible surface of an invisible exchange, where resonance governs the flow of energy and structure from atoms to galaxies. The theorem remains, but the meaning deepens; order as an emergent equilibrium within a resonant field.

26. Newton's Laws of Motion

Newton's laws remain reliable at the human scale, but TOE suggests they rest atop a deeper substrate. Motion isn't just force and mass. It's resonance. Gravity subtly shapes matter through inward waves, making classical laws an emergent approximation of a more complex, dynamic geometry.

27. Koide Formula

Three lepton masses in near-perfect balance. Too precise to ignore. The TOE views this not as coincidence, but as harmonic tuning; resonance in the quantum field crystallising into form. A musical interval encoded in mass itself.

28. Strong Nuclear Force

The nucleus is held together not just by force, but by fit. TOE sees strong interactions as stabilisations within a deeper gravitational resonance. Particles aren't forced to bind; they resonate into coherence.

29. Weak Nuclear Force

The weak force changes particles, shifts identities. TOE suggests that this instability is not a weakness but rather a sensitivity, a responsiveness to shifts in the quantum vacuum, shaped by gravitational fields. Decay becomes transformation guided by resonance.

30. Collapse of the Wave Function

Collapse seems abrupt, arbitrary. TOE proposes a quieter cause. Gravitational resonance nudges quantum states toward coherence.

What appears to be randomness may be the field resolving tension, not breaking symmetry, but finding it.

31. Planck's Constant

Planck's constant defines the quantum grain of reality. But what if it's more than a unit? TOE frames it as a beat, an imprint of deeper resonance; a pulse echoing from gravitational light waves that shape the vacuum itself.

32. Poisson's Equation

It maps how potentials spread through space. Classically, this seems mechanical. But in the TOE, space is alive with inward gravitational waves. Poisson's equation becomes not just a formula, but a trace, a faint imprint of how resonance sculpts the field beneath the visible.

33. Loop Quantum Gravity (Spin Networks)

Space as discrete loops once felt sterile. But reimagined through resonance, spin networks become chords, not scaffolds. Geometry becomes song: quantisation, not a grid, but a rhythm. Gravity, not pulling, but shaping through compression and release.

34. Geophysical Models

Mantle flow, tectonic drift, core dynamics. We treat them as mechanical processes, but TOE suggests they may echo quantum structure. Earth's rhythms could be gravitational harmonics scaled up, the macro mimicking the micro in deep, resonant symmetry.

35. Supersymmetry (SUSY)

SUSY proposes elegance, a symmetry of particles. But perhaps it is not fundamental. TOE frames it as resonance alignment under specific conditions, not a universal truth. The hierarchy problem may dissolve if SUSY is an emergent resonance, not a buried law.

36. Quantum Entanglement

Entanglement seems like instant connection. But TOE suggests it is resonance, two particles locked into a shared structure across dimensions. Not signal, not force, but coherence within an unseen field. The universe does not pass messages. It stays in tune.

37. Non-Equilibrium Thermodynamics

Thermodynamics often assumes equilibrium, yet life, stars, and galaxies emerge from imbalance. The TOE reframes this imbalance not as disorder, but as directed resonance. Gravitational waves may guide entropy, shaping complexity where chaos should reign. Entropy, then, becomes sculptor, not destroyer.

38. Quantum Gravity Theories

Attempts to unify gravity and quantum mechanics have grown dense. TOE offers a simpler view: gravity as an inward, resonant wave, not a force but a rhythm shaping space from within. This soft geometry may bridge the quantum gap without requiring a new mathematical edifice.

39. Gauge Symmetry and Anomalies

We treat symmetry as sacred, yet nature allows it to bend. TOE suggests gravitational resonance might be the cause, subtly tilting quantum fields. What we call anomalies could be signs of this deeper modulation, not errors, but expressions of a more fluid symmetry.

40. Noether's Theorem

Symmetry leads to conservation, but TOE suggests it may also arise from resonance. Inward gravitational waves could generate conserved behaviours not through static laws, but through evolving geometry. Conservation, in this light, becomes a side effect of deeper structure.

41. Holographic Principle

The idea that reality is a projection already challenges intuition. But TOE pushes further. Rather than a flat boundary encoding surface data, it suggests the projection is shaped by gravitational light waves flowing through a higher-dimensional field. Our three-dimensional world may be a standing wave, a resonant shadow of something deeper beyond the veil.

42. Mach's Principle

The idea that your inertia connects to all matter in the universe is quietly profound. TOE reframes this through quantum resonance. Inertia may not be local, but relational, a response to subtle gravitational links threading through space. You move, and the universe adjusts. Or perhaps it moves first, and you follow.

43. AdS/CFT Correspondence

This duality between gravity in a higher-dimensional space and quantum fields on a boundary hints at something deeper. TOE proposes that the bridge is not only mathematical, but also resonant. The bulk becomes a field of oscillations, and what we see at the boundary is its echo. The correspondence, then, is not just structural; it's harmonic.

44. Information Paradox in Black Holes

Black holes are often seen as the end of information. But TOE offers another view. Gravitational waves beneath the horizon may carry and preserve patterns, transforming rather than erasing them. Information does not vanish; it shifts form, encoded in the resonant geometry of quantum space.

45. Cosmic Censorship Hypothesis

The idea that nature hides singularities behind event horizons offers comfort. TOE suggests the veil may not be silence, but structure. Gravitational resonance could redistribute energy before extremes are exposed. When collapse nears, spacetime may not break; it may harmonise. The singularity stays hidden, not by force, but by the rhythm of the field itself.

46. Renormalisation Group in Quantum Field Theory

Infinities in quantum theory are patched with clever fixes. TOE questions whether they should exist at all. If gravitational resonance smooths the field from the outset, then the apparent divergences may dissolve naturally. Renormalisation becomes a sign of deeper order, not a workaround, but a glimpse of a system already regulating itself.

47. Anthropic Principle

The universe appears fine-tuned, but why? The Anthropic Principle offers no real mechanism. TOE suggests resonance may be the missing piece. Gravitational waves, shaped across dimensions, could be sculpting the stage on which life arises. Not intention, not randomness, but alignment. A pattern that allows complexity to emerge, not for us, but through us.

List of Some Unexplained Observational Phenomena TOE Should Help Explain

1. Dark Matter and Dark Energy

Perhaps the mystery isn't in what we haven't seen, but in how we've chosen to look. For decades, dark matter and dark energy have been treated as ghostly scaffolds, patching holes in models that otherwise wobble. But TOE takes another path. It suggests these shadows aren't separate entities at all, but signatures—gravitational resonance embedded in the quantum fabric.

Picture an inward-folding phenomenon, not a classical force, but a compression in space shaped by higher-dimensional waves. Not 'dark' but misread. If gravity acts as a recursive, resonant phenomenon, then many of the universe's so-called anomalies may simply be natural consequences of that deeper structure.

Take cosmic expansion. Instead of a repulsive dark energy inflating spacetime, the TOE proposes gravitational compression and release: shifts in quantum density that distort light, time, and structure. Rotation curves. Lensing. Missing mass. All may be light responding to underlying gravitational phenomena.

We've been watching the surface. The real story may be playing in the folds.

"Dark matter did it" is starting to sound like the cosmological equivalent of "The dog ate my homework."

2. Formation of "Universe Breaker" Galaxies and Quantum Space Halos in the Early Universe

Sometimes anomalies are not anomalies. They are reminders that our models may be built on assumptions that are too rigid, too flat. The discovery of immense, mature galaxies early in the universe's timeline has quietly shaken the foundations of cosmology. These so-called "universe breakers" don't just defy expectation; they undermine the accepted chronology. However, TOE does not require exotic particles or last-minute fixes. It offers a lens shift: gravitational resonance acting on a compressed quantum stage.

In those primordial moments, when energy density was raw and undifferentiated, resonance was not subtle. It was formative. TOE suggests that matter did not coalesce slowly over time, but crystallised rapidly through gravitational compression. Space was not a static container. It was pliable, shaped by waves that folded and pressed it into coherent form. Galaxies may have emerged through symmetry-breaking, not gradual assembly.

As for the halos, those inferred rings of dark matter- TOE reframes them. They are not particle clouds but structural geometries, compressions in the quantum field. These are not scaffolds of missing mass, but contours of curved space. Tensioned folds that shape and stabilise galactic structure.

Gravitational lensing, then, is not the shadow of invisible matter but the response of light to regions of intensified resonance. Uniform rotation curves are not errors in mass accounting, but echoes of how quantum space itself has been modulated.

These galaxies are not outliers. They are signatures. They reflect a universe that formed through pressure and harmony, not chaos and drift.

The real question may never have been how they formed so fast, but why we assumed they had to form slowly.

3. Cosmic Expansion

We tend to picture the universe's growth as a smooth ballooning, a uniform stretch across all directions. But TOE offers a subtler picture. Expansion may not be constant or even global. It may ripple. Gravitational resonance, interacting with the flexible geometry of quantum space, could locally modulate expansion. Galaxies are not simply drifting apart in a passive spread, but moving in response to waves, pulled and released by deep structures beneath the fabric. Space doesn't just stretch, it surges and slackens. Expansion becomes less a law, more a breathing, responsive process.

4. Cosmic Voids and Supervoids

Why are some regions so profoundly empty? Standard models call them voids and move on. TOE sees them differently. In any resonating field, there are nodes: zones of cancellation, quiet, absence. Cosmic voids may be those nodes. Not accidental gaps, but the natural silence within a vibrating field. Gravitational resonance could be drawing mass toward areas of constructive interference, leaving behind low-energy basins where structure cannot form. The emptiness, in this view, is not a flaw. It is the other side of structure.

5. Absolute Rotation

Why do galaxies spin so consistently, so precisely, across such vast scales? TOE proposes it's not just leftover momentum. Once a structure forms, it may lock into the resonance pattern of the space it occupies, like a gyroscope syncing with a deeper rhythm. The inward flow of gravitational light waves doesn't just bind, it tunes. Rotation becomes stabilised by resonance, not merely preserved by inertia.

6. Cosmic Microwave Background (CMB)

Anomalies The Cold Spot in the CMB has resisted tidy explanations. TOE sees it not as an error, but as a quiet region in the gravitational field. A patch where wave density dipped, reducing thermal energy at the time of decoupling. Less resonance. Less heat. 7. Hemispherical Power Asymmetry (HPA) The sky should be uniform. Yet one half carries more structure, more power. TOE suggests directional resonance. Early gravitational waves may have moved with a slight bias, embedding this asymmetry into the quantum substrate. Not noise, but signature. A trace of resonance echoing through the earliest scaffolding of the cosmos.

7. The Cosmological Constant Problem

Vacuum energy should be immense. But what we observe is soft, subdued. The mismatch—over 10^{120} times off—is staggering. TOE offers a different view: vacuum energy is not static, but porous. If gravitational light waves ripple inward through quantum space, their motion could bleed off excess energy into higher dimensions. A kind of quiet release. What looks like a crisis in conventional models becomes, here, an emergent equilibrium. No desperate fine-tuning. Just a vacuum that moves, responds, and breathes.

8. The Koide Formula: A Mathematical Enigma in Particle Masses

Some patterns whisper through simplicity. The Koide formula is one of them. A near-perfect relationship between the masses of the electron, muon, and tau. Three leptons, one ratio. Too precise to dismiss, too mysterious to explain.

TOE takes it seriously. What if mass is not fixed, but emergent? A resonance formed by the interaction of quantum fields and gravitational light waves. In this view, Koide's precision becomes a signature, not an anomaly. A harmonic, stabilised by geometry.

Masses wouldn't be assigned. They would crystallise. Like standing waves on a vibrating surface, lepton masses might reflect deeper spacetime constraints. Gravitational resonance could shape these outcomes, embedding hierarchy not through force, but through pattern.

Conventional physics separates gravity from mass generation. TOE does not. If resonance weaves them together, then Koide might be more than a coincidence. It may be a clue. A fingerprint left by the deeper structure of reality. If TOE is right, then Koide may yet become the Rosetta stone of a new kind of physics.

9. The Fine-Structure Constant (1/137)

And then there's alpha. The fine-structure constant. A pure number, dimensionless, immune to the units we choose. It defines the strength of electromagnetic interaction, stitches quantum electrodynamics together, and yet, why this particular value? Why 1/137?

TOE suggests it's not arbitrary that it's not a constant at all, but a resonance. A value that emerges, naturally, from the interaction between gravitational light waves and the quantum vacuum. Like the pitch of a bell formed by its shape, or the pattern of frost on a windowpane, this number may simply be what the universe is when its deeper symmetries are left to evolve. Geometry becomes law. Vibration becomes constant.

Think of it like this: Fibonacci spirals appear in sunflower seeds, not because someone imposed them, but because they're the natural solution to a packing problem. Alpha might be similar. Not placed there, not imposed, but inevitable. The resonance of spacetime at its most fundamental scale.

And if that's true, then physics is less about discovering static rules and more about listening, carefully, to the harmonies of existence.

Not merely equations, but chords. Not just constants, but frequencies echoed through gravitational light and quantum foam.

10. Subatomic Shifts and the Fluidity of Time

Time, as we know it, appears constant. A steady tick, measured by atomic clocks, unwavering and absolute. But at quantum scales, where gravitational resonance meets the flicker of particle fields, that steadiness may falter.

Atomic clocks, though engineered for precision, could be quietly distorted by subtle quantum fluctuations. These aren't glitches, but signals. Indications that time's apparent uniformity may be shaped by deeper, less stable forces.

If gravitational light waves are influencing the quantum substrate, then time might be emergent, not fundamental. Less a fixed flow, more a resonance itself; arising from, and shaped by, invisible patterns in the quantum fabric. If time can bend this quietly, what else might be slipping through the gaps?

11. Gravity's Whisper in the Quantum Realm

What if gravity isn't only shaping galaxies and black holes, but also influencing the quantum world? Not as a loud, prominent force, but as a subtle pressure- gravitational waves threading through spacetime, gently nudging quantum states.

For decades, quantum mechanics has rested on the idea of pure unpredictability. But perhaps that randomness isn't absolute. Maybe near massive objects, or in the early universe, gravity leaves a faint imprint, tilting probabilities in quiet, measurable ways.

Instead of chaos at the smallest scales, there may be a hidden order; gravity whispering into the quantum domain. Not overpowering, but present. Not dictating outcomes, but shaping the field in which choices unfold.

2. The Inward Collapse of Stars and the Alchemy of Rotation We tend to focus on the brute collapse of mass when stars die, watching them cave into singularities. But TOE invites a closer look. Gravitational light waves may be the engine behind this collapse, compressing matter not just inward, but into a new kind of order. Rotation, far from incidental, could act as a resonant driver, sustain-

ing and shaping this process. As mass tightens, space, energy, and matter may begin to blur. Singularity might not mean breakdown. It might signal transformation. A zone where structure liquefies and resonance takes over.

13. The Fragility of a False Vacuum

The false vacuum is often imagined as a silent cosmic threat, a sudden shift into a deeper energy state. But TOE reframes it. What if vacuum decay isn't a dramatic rupture, but a gentle and ongoing process? The fabric of quantum space may already be in constant fluctuation, with gravitational resonance nudging the field toward lower-energy states. Not a catastrophic reset, but a quiet, continuous tuning. The vacuum isn't stable: it's adaptive, responding to waves and pressures that slowly reshape the universe from within.

14. The Quantum Signature of Life: Fluorescence and Beyond

What if life isn't just a lucky arrangement of molecules, but a direct product of quantum resonance? Take fluorescence- the glow from certain molecules when excited by light. Usually dismissed as a minor detail in biology, TOE suggests it could point to something deeper. A quantum interaction. A signal that life is tuned to the patterns of the universe. Perhaps our biology isn't just the result of random mutation, but a natural response to resonant structures woven through quantum space. In this light, life becomes less an accident, more a harmony. A structure aligned with the fabric of reality itself.

15. Mass Gains in High-Velocity Particles: The Gravitational Hand at Play

Special relativity provides the framework for understanding mass increase at high velocities, but could the underlying mechanism be more gravitational in nature? If gravitational light waves interact with high-speed particles, subtly compressing quantum space around them, then mass gain may not be merely a relativistic artefact but an emergent property of a deeper gravitational interplay. Intriguingly, while velocity shifts mass, the rest mass itself may remain untouched, suggesting that gravity is not merely a force but perhaps the unseen arbiter of motion.

16. Euler's Formula, Fibonacci Spirals, and the Blueprint of Reality

From the spiral arms of galaxies to the branching of veins in a leaf, nature echoes with geometric motifs. Fibonacci sequences, golden ratios, self-replicating spirals. But why? The answer may lie not in arbitrary aesthetics, but in the resonant harmonics of quantum space itself. If underlying gravitational structures shape matter at its most fundamental level, then these recurring patterns are not mere coincidences but the inevitable fingerprints of a deeper order, one that whispers its presence in both the celestial and the cellular.

17. Strong and Weak Nuclear Forces Gravitational Resonance, Quark Confinement, and the Unfinished Puzzle of Unity

We tend to compartmentalise the forces of nature - strong, weak, electromagnetic, gravitational - as if they exist in tidy isolation. But that framing starts to fracture at the subatomic scale, where certainty thins and familiar assumptions begin to flicker.

Start with the strong force. It's immense, yet oddly local. It binds quarks into hadrons, but no one has ever isolated a single quark. The more you try to pull them apart, the more force you unleash, until the system births new quarks instead. That's confinement. A rubber band stretched to breaking, only it never snaps.

Resonating Waves in QCD But what if confinement isn't just a story of gluons and colour charge? What if gravity, not the Newtonian kind, not even Einstein's curvature, but a deeper, fourth-dimensional resonance, has a role here too? A wave-like force that doesn't just respond to mass but intensifies and reverberates across quantum space.

Gluons are said to weave flux tubes between quarks. But what if those tubes are not just QCD artefacts? What if gravitational resonance helps shape their geometry? Not as a background hum, but as an active player.

Hadron Structure We already know that hadronic models start to wobble at extreme energies, inside neutron stars, or in the first moments after the Big Bang. In these regimes, gravity and the strong force may blur. Quark-gluon plasma, that so-called primordial soup, might not just be a thermal artefact, but a state triggered by gravitational resonance itself.

Strong Force Beyond the Standard Model

Now, consider the weak force. Less headline-grabbing, but no less strange. It governs radioactive decay, has already merged partly with electromagnetism, yet hides one of the deepest mysteries in physics: why the universe leans toward matter.

Unification with Electromagnetic Force (Weak Force), Electroweak theory doesn't explain why symmetry breaks in favour of matter. But if gravitational resonance subtly interacts with the weak force, if those waves amplify matter over antimatter. Then CP violation might not be a bug, but a built-in asymmetry. Not random, but guided.

CP Violation (Weak Force) Gravitational resonance might explain why the universe chose matter, not as a roll of the dice, but as a result of wave interference patterns embedded in the quantum vacuum. This shift in perspective could provide a coherent mechanism for one of cosmology's oldest puzzles.

Precision Tests and Anomalies

Then come the anomalies. Those stray data points in high-precision tests that don't quite fit. Perhaps they're not noise, but resonance. Subtle shifts in measurement caused by a gravitational interaction that the Standard Model was never built to detect.

Potential Resolutions from TOE If TOE is even close, then unification may not require a zoo of particles or billion-dollar colliders. It might come from reframing what we already know. Gravity wouldn't just be a background force; it would be in dialogue with the strong and weak interactions. A quiet unification, detectable not through speculation, but through high-energy physics and cosmological data already within reach.

It doesn't claim to be complete. But it offers a path. And that path suggests that the strong and weak forces may not be isolated at all. They may already be resonating with gravity.

18. Strong CP Problem and the Gravitational Resonant Wave

1. The Curious Case of the Strong CP Problem In Quantum Chromodynamics, there's an odd silence where noise should be. The strong force, which binds quarks into protons and neutrons, seems to preserve CP symmetry far more than expected. Theories say it should violate this symmetry. Yet experiments, like those searching for the neutron electric dipole moment, find nothing. It's not just unexpected; it's suspicious.

2. Gravitational Resonance as the Missing Piece. We assume gravity is too weak to matter in the quantum world. But what if it's not? Not in a dramatic way, just enough to shift the balance. If gravity carries a resonant, wave-like component. Subtle, dimensional, pervasive, it could be suppressing those violations. Not by force, but by favouring balance. A quiet interference. A harmonic that doesn't allow the strong force to tip.

3. How This Suppression Might Work

Perhaps gravity acts like a filter. Not blocking CP violation, but softening it; quenching it before it manifests. Gluons don't operate in isolation. They ripple through a quantum soup. If gravity resonates through that same medium, it may be tilting the field, muting asymmetries before they rise. Think noise-cancelling, not erasure, just perfect opposition. The rulebook remains, but one of the lines is never read.

4. The Neutron Electric Dipole Moment. A Hidden Signal? We keep listening for a dipole moment. Nothing answers. But perhaps it's not silence; it's suppression. TOE suggests gravitational resonance may be dampening the signal, pushing it below our threshold of detection. Not absence, but masking. The moment may be there, faint and filtered, drowned in the hum of a deeper geometry.

5. New Predictions and Hidden Consequences If this view holds, future experiments might notice small deviations, not from new particles, but from familiar ones behaving oddly in extreme conditions. In neutron stars or high-energy collisions, gravitational resonance might reveal itself as tiny misalignments in expected outcomes. Not fireworks, but fine print. TOE hints that the strong force may already

be entangled with gravity, and the universe's symmetry is held, not by silence, but by song.

19. Black Holes in the TOE Framework: Anchors of Galactic Structure and Catalysts of Quantum Space

Dynamic Influence and the Genesis of Quantum Space Stephen Hawking's insight that black holes radiate was revolutionary. These were no longer cosmic vacuums, but dissipative systems, shedding mass through quantum effects. Yet TOE takes it further. In this framework, black holes are not just endpoints. They are active participants in the evolution of quantum space. Their immense mass generates gravitational resonance that reshapes the surrounding quantum field. Not metaphors, but real disturbances, forming and maintaining the geometry of space.

Black Holes as Galactic Anchors and Resonant Cores In TOE, black holes are more than spacetime scars. They are structural centres. Galaxies orbit them not just through gravity, but in response to the resonant fields they emit. Rotational coherence in galactic disks, often attributed to dark matter, may instead reflect dense quantum geometry, shaped and amplified by gravitational resonance. These black holes organise matter through pressure waves, stabilising complexity. What we call missing mass may be a misunderstanding of this intensified geometry.

Quantum Space and Cosmic Expansion If black holes structure galaxies, they may also influence cosmic expansion. In this view, space grows not through uniform stretching, but through textured resonance. Black holes vibrate the quantum field outward, expanding space by agitation, not by inflation. This adds direction and rhythm to expansion, seeded by the churn of resonant cores spread across the universe.

The Black Hole as a Cosmic Furnace

Rather than perfect absorbers, black holes in TOE are transformative.

Matter crossing the event horizon undergoes a phase shift. What returns as Hawking radiation may not be heat alone, but rather restructured information, a form of feedback from within. Black holes process entropy into resonance, converting structure into energy and releasing it in subtle, encoded forms. They act as engines, not endpoints.

Internal Structure and Gravitational Resonance

Conventional models fail inside a black hole. TOE offers a different view. Collapse doesn't continue into a singularity. Resonance builds to a point of balance. The mass stabilises, forming a dense, perhaps geometric core. No infinite density, just resonance at maximum. This reframes black holes as regions where gravity organises, rather than destroys. Not breakdowns, but condensations of order.

Gravitational light waves, central to TOE, are not analogies. They are agents. Near black holes, they compress and structure the quantum medium, stirring it into coherent formations. The black hole becomes a point of convergence, where light, geometry, and quantum fields meet. It doesn't just distort reality, it refines it.

The Information Loss Paradox Revisited

The long-standing puzzle of whether black holes erase information is reinterpreted here. Information is not lost. It transforms. As mass falls inward, resonance encodes its pattern into the gravitational field. The event horizon becomes a membrane, not a wall. Hawking radiation is reimagined as a slow transmission of information, encoded and released through resonance over time.

Entropy as Resonant Complexity Entropy, in TOE, becomes more than surface area. It reflects the complexity encoded by gravitational compression. In black holes, quantum space becomes dense with layered information. The Bekenstein-Hawking formula is not wrong, but incomplete. It misses the resonance. Information is not smeared thin: it is stacked and sculpted, with depth tied to gravitational pressure.

Final Thoughts: Black Holes as Cosmic Architects

In this model, black holes are not anomalies. They are key. They structure galaxies, regulate entropy, and help shape the evolution of space itself. They are not points of failure for physics, but intersections where gravity and quantum coherence align. They do not erase. They reprocess. They do not just end stars, they set the stage for what comes next.

Black holes are not cosmic graveyards. They are the engine rooms of the universe.

20. The g-2 Anomaly

The g-2 anomaly isn't just a statistical hiccup. It's a crack. A signal that something beneath the surface may be misaligned. Gravity, in the TOE framework, isn't background curvature- it's pressure. A resonant wave folding inward, tunnelling through quantum space.

What happens when such a wave brushes up against a particle like the muon, already perched at the edge of quantum instability? Heavier than the electron, strangely reactive, it sits in a turbulent sea of fluctuations. A gravitational resonance might gently nudge that sea, just enough to shift its magnetic moment. Not a new force. Not a new particle. Just gravity, redefined. Expressed not through mass or curvature, but through resonance.

The anomaly may not be a glitch in data or a triumph of discovery. It may simply be a misreading: one caused by mistaking gravity's true nature.

21. Cosmic Filaments

Cosmic filaments are too clean, too taut to be mere chance. They stretch across intergalactic space like drawn threads, structured and precise. The standard model says they're the result of matter collapsing under gravity. But that leaves something out.

In TOE, these filaments are shaped by gravitational resonance. Picture an inward-folding pressure guiding matter through the quantum substrate, aligning it where the tension flows most cleanly. Not gravity pulling, but resonance sculpting.

They aren't just structures, they're harmonics. Visible traces of a deeper pattern written into space itself. Not formed by random clumping, but by organised interference. They are the echo of a larger geometry, one that doesn't obey the conventional divide between gravity and quantum mechanics.

22. Planet 9 and the Quantum Space Compression Hypothesis

Planet 9 captures the imagination; a hidden world beyond Neptune, silently pulling at icy bodies on the solar system's edge. But maybe we've misunderstood the clues. Maybe the anomaly isn't caused by a massive object at all, but by subtle compression in the quantum fabric of space itself.

In the TOE framework, space is not empty or passive. It is an emergent medium, shaped by inward gravitational resonance. Unlike classical spacetime, which curves under mass, quantum space may compress in response to deeper structural pressures. This compression could influence the motion of distant objects without requiring any additional matter.

The clustering of trans-Neptunian orbits might be a symptom of this effect. Not the result of a rogue planet, but of regional space compression. Like taut fabric pulling gently inward, creating measurable shifts without visible cause.

Think of ripples on a pond. Not all are caused by objects striking the surface. Some come from below, stirred by hidden forces. Similarly, distortions in orbital paths may arise from gravitational pressure within quantum space, rather than from unseen planetary mass.

This could explain why Planet 9 remains undetected. Our instruments search for light and mass. But if the anomaly is a structural compression, not a luminous object, it would leave no shadow. Just a quiet signal embedded in motion.

Planet 9 may be a placeholder for an assumption: that gravity always originates from mass. But TOE redefines gravity as resonance. Compression in the quantum field, driven by gravitational intensity, could create the same effects we attribute to matter. No missing planet, just a deeper architecture we have not yet fully recognised.

Chapter 16

The Theory of Everything V

23. Earthquake Lights and Ball Lightning Before Seismic Events

The eerie glow of earthquake lights: those strange orbs and flashes seen near seismic zones have long defied tidy explanation. Traditional theories suggest that plasma or electric fields originate from tectonic stress, but the phenomenon remains too erratic and elusive. Perhaps we've misread them entirely.

From the TOE perspective, these lights may be more than surface-level electrical events. As stress builds in fault lines rich in piezoelectric minerals like quartz, pressure-induced discharges occur. But if gravitational resonance is also moving through compressed quantum space at these moments, the result could be more than just electrical sparks. The crust may act as an amplifier, not just of pressure, but of spacetime fluctuations, lighting up the air in ways that straddle categories.

These emissions might not come from above, but from below. Not purely electromagnetic. Not entirely seismic. They could be visual echoes of gravitational waves interacting with matter, fleeting artefacts of a deeper resonance crossing the boundary between Earth and space.

If so, the Earth isn't just a mechanical body. It's responsive. Tectonic stress might occasionally crack the surface of spacetime itself, revealing a subtle, luminous trace of gravity's hidden structure.

24. Anomalies in the SETI Catalogue

The SETI archive contains signals and patterns that resist easy explanation: flashes of radiation, irregular pulses, and astronomical arrangements that, under classical physics, shouldn't exist. One of the most puzzling is the so-called impossible triple star system. According to standard models, three stars in close orbit should destabilise quickly. Too many gravitational pulls. Too much chaos. Yet these systems remain stable over millennia.

What holds them together? The Theory of Everything offers an alternative view. If gravity is not merely a force but a resonant, inward-pulling wave in higher-dimensional space, then perhaps these stars are held in place through resonance. Their gravitational fields may sync like harmonic tones, compressing the quantum space between them. Instead of clashing, they stabilise, amplifying and balancing through shared rhythm.

Other anomalies arise from sudden, powerful flashes, such as gamma-ray bursts. Many fit known patterns: collapsing stars, merging black holes. But some do not. Their placements seem odd, their timing peculiar. They appear patterned, even intentional. Some speculate they are signals: messages from intelligent life, traces of ancient minds leaving markers in the dark. But perhaps the story is different.

What if these bursts aren't artificial? What if they are natural outcomes of resonance itself? Gravitational waves peaking through quantum space, triggering brief but intense releases of energy. Not messages, but echoes. Not intelligence, but geometry.

These signals may be real. The mystery may not lie in their presence, but in our expectations. We might be asking the wrong question.

25. Neutrino Anomalies in Earth's Ice: Insights from ANITA and Gravitational Resonance

1. ANITA's Findings and Quantum Resonance in Ice The ANITA experiment, conducted in Antarctica, detected anomalous signals that challenged conventional understanding. Neutrinos, typically elusive and indifferent to matter, appeared to interact in unexpected ways. But what if these signals weren't caused by neutrinos at all?

In the Theory of Everything (TOE), such detections may arise from gravitational resonant waves interacting with the structured environment of ice, creating quantum effects that mimic neutrino behaviour. Ice, with its crystalline structure, offers an ideal medium. As gravitational waves pass through, they could induce fluctuations--ripples in the quantum field--that resemble particle interactions but are actually gravitational in origin.

In this view, the signals aren't from passing neutrinos. They emerge from the interplay between gravitational resonance and quantum space, shaped by the unique geometry of the ice.

2. Natural Quantum Resonance Structures in Ice and Saturn's Rings TOE draws a parallel between Earth's polar ice and Saturn's rings. Both environments exhibit resonant properties that may influence particle-like interactions in unexpected ways. In Antarctica's ice, quantum resonance fields could redirect or mimic the paths of particles, generating signals we interpret as neutrino events without the particles themselves.

Saturn's rings, with their highly ordered structure, may reflect the same phenomenon on a cosmic scale. Gravitational resonance interacting with quantum matter could shape large-scale structures as well as localised quantum fields. Just as Saturn's rings are shaped by gravitational forces acting on ice, so might Earth's ice sheets host structured resonance fields capable of triggering detectable anomalies.

Recent studies suggesting doughnut-shaped quantum formations within Earth's core add further context. These structures suggest that gravitational forces are shaping resonant quantum fields deep within the planet, a phenomenon that may also be active in surface environments, such as polar ice. ANITA's detections, then, could be byproducts of these resonant interactions rather than evidence of exotic particles.

TOE reimagines the anomalous signals not as breakthroughs in particle detection, but as signs of gravitational resonance shaping the quantum field in specific materials. By reframing the events through this lens, the theory offers a coherent explanation that aligns with broader cosmic and terrestrial patterns.

Additional Anomalies that the Theory of Everything (TOE) could help to explain

1. Chiral Anomaly

The subtle imbalance between left-handed and right-handed particles has long been a puzzle. It may not be an arbitrary asymmetry, but a result of gravitational light waves inducing a resonance that subtly biases certain quantum states. This influence could skew chirality at a fundamental level, embedding asymmetry into the fabric of the quantum world.

2. Conformal Anomaly

In theory, the laws of physics should remain unchanged under a shift in scale. But in practice, scale invariance breaks down. The TOE suggests this may be a product of gravitational resonance, where 3D light and deeper 4th-dimensional gravitational fields interact. These resonant interferences may distort the local energy landscape, making it impossible for quantum fields to maintain perfect scaling symmetry. Where resonance is absent or out of phase, anomalies emerge.

3. Gauge Anomaly

Gauge symmetries are meant to safeguard consistency in the mathematical structure of particle physics. But under certain conditions, they falter. TOE proposes that this isn't a flaw in the theory, but an overlooked influence: gravitational light resonating with quantum fields. These interactions may subtly deform the geometry of symmetry, producing the anomalies that physicists have long struggled to reconcile.

4. Global Anomaly

Some inconsistencies seem to persist across entire systems, transcending local descriptions. TOE suggests that if 4th-dimensional gravitational waves modulate quantum space globally, rather than locally, then small distortions may cascade across large-scale transformations. Global anomalies might not be errors in calculation, but rather consequences of a deeper, ever-present resonance at work behind the scenes.

5. Gravitational Anomaly

General relativity insists on diffeomorphism invariance, yet certain scenarios break that rule. TOE proposes that gravitational light waves, especially when resonating through spacetime at extreme densities, might temporarily override that symmetry. This dynamic reshaping of geometry, while subtle, could provide a rare window into where gravity and quantum theory collide.

6. Mixed Anomaly

When gravitational and electromagnetic fields overlap, the result is rarely straightforward. According to TOE, the intersection of 3D and 4D light waves may create resonant cross-effects that challenge traditional symmetries. These mixed anomalies aren't necessarily contradictions. They may be signs of deeper field interactions we've yet to map fully

7. Parity Anomaly

Why does the universe prefer certain orientations? Parity violation is one of nature's quiet riddles. TOE suggests that gravitational light waves, with their inward-pulling influence, may not act symmetrically. If quantum fluctuations are nudged more easily in one direction than another, parity could be broken, not by chance, but by resonance.

Types of celestial bodies and phenomena the Theory of Everything (TOE) could help to explain.

1. The Sun's Corona: Quantum Compression and Decayless Oscillations

One of the great mysteries of solar physics lies just above the Sun's visible surface. The corona, despite being farther from the core, reaches temperatures of several million degrees, far hotter than the surface below. Traditional models invoke magnetic reconnection and particle acceleration to explain this, but something remains incomplete.

The TOE suggests an alternative: quantum space compression near massive objects, particularly those with high rotational and magnetic dynamics, could trigger resonance effects. In this view, the corona's heat is not an incidental effect of magnetism but a symptom of deeper resonant phenomena, gravitational waves stimulating energy transfer at the quantum level.

Even more intriguing is the presence of decayless kink oscillations within the Sun's coronal loops, oscillations that, oddly, do not dissipate over time. These enduring waveforms suggest a constant energy input from a source that has not yet been clearly identified. TOE proposes that this source might be gravitational resonance itself, a standing wave feeding energy back into the loop, echoing through space like a plucked string resonating with invisible geometry.

Recent three-dimensional polarisation analyses of these solar waves offer clues. Their structured forms hint at geometric underpinnings, resonant properties that may be universal, governing not only stars but the particles and forces we encounter at far smaller scales.

2. Earth's Core and Doughnut-Shaped Quantum Structures

Moving inward from the macrocosmic to the planetary, Earth's interior holds its own set of mysteries. While heat in the core is typically attributed to pressure and radioactive decay, these explanations fall short of accounting for certain thermodynamic and electromagnetic anomalies.

The Theory of Everything proposes that quantum resonance may contribute here as well. Gravitational resonant waves, strongest near dense mass, could amplify energy exchanges within Earth's core. These interactions may generate not only heat, but also subtle patterns in electromagnetic behaviour and complex field dynamics.

Recent discoveries of doughnut-shaped quantum structures near Earth's centre support this view. These toroidal patterns, formed under immense gravitational compression, may act as resonant cavities, channelling gravitational light in ways that influence molten iron flow and magnetic polarity.

Gravitational Resonant Light Waves and Layered Planetary Interactions According to TOE, gravity operates as a fourth-dimensional light wave, compressing quantum space and interacting with matter differently depending on density, temperature, and structure. Earth, like many celestial bodies, is stratified into layers that may respond uniquely to gravitational resonance.

1. Intense Resonance in the Core

The core, as the most compressed region, is where resonance effects are likely strongest. Doughnut-shaped quantum structures may generate standing wave fields that modulate molten flow and feed back into Earth's geomagnetic system.

2. Magnetic Field Reversals and Resonance Cycles

Earth's magnetic field reverses polarity in cycles that appear irregular, sometimes as short as 7,000 years. TOE suggests these reversals may be driven by internal resonance cycles. As gravitational resonance intensifies or wanes, it could reorganise flow within the core, shifting the magnetic axis.

3. Tectonic Activity and Geological Evolution Resonance may also influence tectonic processes. Variations in mass across the mantle could lead to asymmetric gravitational interactions, subtly guiding tectonic drift, volcanic patterns, and seismic behaviour. Over geological timescales, these forces may shape continents, rift zones, and ocean basins. Not just as mechanical consequences, but as expressions of underlying resonant geometries driving Earth's internal motion.

3. The Enigmatic Magnetic Fields of Neptune and Uranus

Neptune and Uranus, cold and distant ice giants, remain among the least understood planets in our solar system. Despite the name, they are not composed entirely of ice. Their atmospheres are turbulent blends of hydrogen, helium, and methane, while below lies a mantle of ammonia and water ices surrounding dense rocky cores.

But it is their magnetic fields that defy expectations. Unlike Earth's relatively aligned dipole, the magnetic axes of Neptune and Uranus are sharply tilted and significantly offset from their planetary centres. Viewed through the lens of gravitational resonance, one possible explanation is that their layered, non-uniform interiors enable quantum interference between gravitational waves and conductive materials.

While planets like Jupiter have more fluid, uniform interiors that distribute energy evenly, Neptune and Uranus contain icy, stratified mantles where resonance effects may become chaotic.

Instead of stabilising the magnetic field, these icy interiors could amplify or distort resonant interactions, giving rise to the unusual multipolar magnetic patterns observed. Rather than anomalies, these configurations might reflect the deeper influence of resonance interacting with layered mass and electrical conductivity.

This points to a more complex relationship between internal composition, gravitational resonance, and magnetic structure, one not yet fully captured in traditional planetary models.

4. Mars' Axial Wobble and the Influence of Inward Gravitational Waves

Mars is often regarded as a planetary sibling to Earth, one that diverged from its development early on. In contrast, Earth maintains a stable axial tilt and a strong magnetic shield; Mars wobbles. Its axial precession: a slow, rhythmic shift in its rotational alignment- hints at deeper interactions between its interior and the gravitational forces acting upon it.

Without a large, stabilising moon, Mars is more exposed to both internal and external perturbations. Its smaller, partially liquid core may be especially sensitive to the influence of inward gravitational waves. If these waves interact resonantly with the Martian interior, they could subtly shift its core over time, driving long-term changes in the planet's spin.

Axial Precession and Core Interactions

Unlike Earth, Mars lacks a lunar anchor. That absence matters. With no large moon to steady its axis, the Red Planet is more vulnerable to subtle shifts. Its axial precession is not just a mechanical remnant. Something is interacting with the core, gently but persistently.

TOE suggests that inward gravitational waves, as four-dimensional resonant currents, may influence the Martian interior. These currents could reach into the core, nudging it over geological timescales and causing shifts in axial orientation.

Earth, protected by a geomagnetic field and driven by an active dynamo, resists such influence. Mars, lacking these stabilisers, absorbs it.

Resonance and Planetary Realignment

From this perspective, planetary stability is shaped not only by mass and angular momentum but by how deeply a planet is embedded in resonant gravitational fields. Earth expresses these influences through magnetic pole reversals. Mars, with a weaker magnetic field, exhibits axial drift.

This shift in view suggests that planetary evolution and stability, possibly even habitability, may depend on the subtle relationship between mass and gravitational resonance. A deeper architecture may be at work, one that governs more than just orbital mechanics. It may shape the very way planets form and hold themselves together.

5. Gas Giants Beyond Our Solar System

Exoplanetary gas giants, bloated and adrift in the dark, challenge classical astrophysics. These are not just curiosities. Many are larger than Jupiter, hotter, stranger, and situated in locations that defy traditional formation models.

Perhaps it is not the planets that are peculiar, but our framework that is incomplete.

If gravitational resonance is real and not metaphor, then these gas giants may be behaving exactly as they should.

1. Formation Beyond the Snow Line

Standard theory says gas giants form beyond the snow line, where frozen volatiles help accumulate mass. Yet many have been found dangerously close to their stars, where heat should strip away their atmospheres.

TOE suggests that inward gravitational resonance could be shaping and stabilising them. Rather than defying their environment, these planets may be held in place by harmonic pressures we have yet to map.

2. Hot Jupiters

Some orbit in just a few days, yet their atmospheres remain intact. Many appear overinflated and should be unstable, but persist.

Instead of relying on planetary migration as a patch, TOE proposes that internal coherence is maintained by gravitational resonance. The energy is high, but resonance may act as a structural scaffold, preserving form under stress.

3. Puffy Gas Giants

Some gas giants are vastly oversized relative to their mass. This low-density inflation challenges classical explanations.

If inward gravitational waves induce rhythmic oscillations, these waves may be stretching the atmosphere from within. The result would not be instability, but structured expansion. A standing wave in planetary form.

4. Magnetic Fields and Atmospheric Stability

Close-in giants should have lost their atmospheres long ago. Yet many exhibit strong magnetic fields or survive where they should not.

TOE suggests that magnetic fields could emerge from resonant charge flows within. The result is not solely a product of convection, but of gravitational wave interaction with the planet's quantum architecture.

5. Diversity in Mass and Orbit

Gas giants come in wildly different sizes and distances. This variation exceeds what standard formation models comfortably allow.

Resonance offers a guiding influence. If mass accretion and orbital drift are shaped by resonant field intensity, then each planet's path becomes a slow dance, not a collision. Variability is not error but adaptation to deeper field dynamics.

6. Atmospheric Composition

Some gas giants carry unexpected molecular signatures. Methane, ammonia, and water appear in unusual concentrations.

TOE suggests that gravity, resonating through quantum space, may influence how molecules cluster. Subtle gradients could draw elements into patterned zones during formation, shaping composition through rhythm rather than chance.

Conclusion

These gas giants are not anomalies. They are signals. Gravity may be more than force; it may be structure. A rhythm. A scaffold for matter and form.

TOE expands the model. It reframes gravity not as a pull, but as presence. Not just the engine of motion, but the architect of shape and stability.

6. Saturn's Formation and Resonant Gravitational Effects

Saturn stands as a strange and elegant exception, an archetype that challenges the simplicity of classical planetary models. With its weightless rings and hydrogen-rich bulk, this gas giant suggests that something more nuanced, something deeper, is shaping matter beyond the familiar hand of gravity and spin. In the TOE framework, this something is gravitational resonance.

Not merely a force, but a resonant field. A selective organiser. A sculptor not only of motion, but of structure itself.

Formation and Material Distribution

In Saturn's early formation, classical models suggest chaos, dust and gas clumping into a planet through blind accretion. But what we observe is far more structured.

Saturn's immense gravity, coupled with its rapid spin, did not produce a random blend of materials. It yielded a layered system that seems filtered, shaped not by chance, but by resonance.

TOE posits that gravitational resonance acted as a kind of sieve, sorting matter by its responsiveness to gravitational light waves. Lighter elements, such as water, ice, and volatile gases, were pushed outward into orbital bands, unable to match the resonant frequency required to fall inward. Denser elements, less reactive, sank toward the core.

This idea is visible in Saturn's structure. The rings, mostly water ice, orbit above an atmosphere notably lacking in water. Standard models struggle to explain this. TOE provides clarity: water was expelled to the resonant edge, while the planet retained only what could sink through.

Layering and Compression

As Saturn's mass increased, so did the depth of its resonant influence. The planet didn't simply gather matter; it organised it. What stands out is not just the sorting of layers, but the compression patterns within them.

Gravitational waves deepened internal density zones, pulling heavy elements like rock and metallic hydrogen into tightly bound layers. Meanwhile, centrifugal forces from rapid spin flung volatiles outward.

The result: a planet of striking contrast. A dense core wrapped in hydrogen, crowned by a sharp, radiant ring system.

These rings, far from being debris or a broken moon, look like the product of an active filtration process. Resonance determines what stays close and what stabilises in orbit.

The Resonant Filtering Mechanism

Saturn may be our clearest case of gravitational resonance shaping an entire planetary system. It operates less like a chaotic body and more like a harmonic structure, defined by how matter interacts with spacetime itself.

Three key effects emerge

1. Separation: Water and volatiles, more responsive to gravitational light waves, were repelled into stable orbits; denser materials fell inward.

2. Layering: Saturn's rings are formed in discrete zones, not simply through collision dynamics but through oscillatory resonance, much like the hexagonal polar storm, itself a resonant pattern.

3. Compression: Internal structures became increasingly stratified, with core materials compressed into dense layers, while the upper atmosphere remained light and resonant.

This isn't just about Saturn. It's about recognising resonance as a fundamental process in nature, one that organises complexity through selective attraction. If gravitational resonance can structure a planet so elegantly, it may well be a foundational principle of formation across the cosmos.

Saturn, then, is not just a planet with pretty rings. It's a cosmic diagram, a resonant artefact.

7. The Hexagonal Storm on Saturn

At Saturn's north pole swirls a storm. That much is expected. But not its shape. A nearly perfect hexagon, spanning a region twice the size of Earth. It has remained for decades, unmoved in form, its six-sided symmetry steady.

Even stranger, the storm's colour has changed; from a bluish green to a muted gold. Not just chemistry, perhaps, but something more subtle, shifting in the structure of its environment.

A Geometric Anomaly

Hexagons don't usually appear in weather systems, not like this. Meteorologists point to jet streams, shear forces, and turbulence. Some explanations partly hold. But a storm forming and maintaining a hexagon for years? That defies simple weather dynamics.

It begins to resemble not random turbulence, but a pattern: something embedded in the fabric of the planet, revealed by movement rather than caused by it.

Gravitational Resonance in the Atmosphere

If Saturn is shaped by gravitational resonance, that influence would not stop at its core. It could extend outward into the atmosphere.

Imagine gravity not only pulling, but humming. A layered, gravitational light wave interacting with gases above the pole. The storm then becomes a kind of echo chamber, where waves settle into standing patterns.

The hexagon may be the surface trace of a deeper resonance, held in place by frequencies moving through the planet.

The Colour Shift

That golden hue didn't used to be there. Saturn's pole, once cooler-toned, now glows warmer.

Sunlight may explain part of it. But it could also hint at a more profound shift, a change in gravitational behaviour subtly altering the quantum interactions in the atmosphere. If space-time itself wavers in density, even slightly, light may scatter differently. Gravity might shape not only mass, but also how energy moves through gas, affecting how colour appears.

A Glimpse of Cosmic Order

So what are we seeing? A storm? A curiosity?

Or something deeper? A fingerprint of resonance, the same force that filters Saturn's rings and sorts matter by its gravitational response, now visibly etched into the clouds above.

8. Earth's Moon and the Quiet Mechanics of Resonance

The textbook version of the Moon's origin--the giant impact hypothesis--has a certain cinematic appeal. A Mars-sized body collides with early Earth, and from the debris, the Moon emerges. Violent, dramatic, and satisfyingly simple.

But it does not quite hold up. Not entirely.

The Moon's composition, especially its isotopes, is too similar to Earth's. Not just broadly compatible, but nearly indistinguishable. If another world had truly collided with ours, we would expect more variation. More foreign signatures. Yet we do not see them. So perhaps the story is wrong.

Resonance Instead of Catastrophe

What if the Moon were not born from a single chaotic event, but from a gradual unfolding? A slow, resonant emergence rather than an explosion.

In this view, the early Earth, molten, radiant, and unstable, was surrounded by a turbulent halo of matter. Not quite rings, but proto-rings. Material suspended between falling back and breaking free. A liminal region shaped by gravity, heat, and rotation.

Now add resonance.

In such an environment, certain orbits become favoured. Zones where gravitational feedback stabilises particles rather than dispersing them. Regions where matter does not just drift, but lingers. Coalesces. Gathers. And over time, becomes something more. Something moon-like.

A Process of Thermal Filtering

Heat played a crucial role. Early Earth radiated so intensely that nearby matter was subjected to extreme thermal forces. Some materials vaporised. Others cooled and settled.

TOE proposes that resonance is not just about balance, but also sorting. A kind of cosmic sieve. Dense elements might have stayed close, while lighter ones were pushed outward or never captured at all.

This filtering could explain why the Moon resembles Earth's crust more than anything else. It is not alien. It is local. The by-product of Earth's own energy, shaped by gravitational tides and thermal pulses.

Extending the Idea Beyond Earth

If that is true here, could it be true elsewhere?

Perhaps moons do not need collisions to form. Perhaps they emerge naturally when the right parameters align: mass, rotation, heat, and time. TOE proposes that satellites aren't debris of violence, but expressions of balance; formed within the turbulence, not from it.

When we observe the moons of Jupiter, Saturn, or exoplanets circling distant stars, we may begin to see similar patterns.

Not identical, but recognisable.

Moons as Inevitable Outcomes

This reframes the story. Not just the Moon's history, but how we understand planetary systems.

Our Moon, in this view, is not a fluke. Not the random aftermath of cosmic violence. It is a near-inevitable outcome of resonance unfolding over time. Like the rings of Saturn or the stable orbits of planets, it arose from a system seeking coherence.

Not an accident. A pattern.

A quiet, round reflection of the same forces that shape the entire cosmos.

The Formation of the Solar System – Through the Lens of TOE

The standard story is familiar. A swirling disc of gas and dust collapses under gravity, forming the Sun and planets. Collisions shape worlds like cosmic Lego bricks, jostling, snapping into place.

It is tidy enough for textbooks, but start looking closely, and something feels off.

Not just the spacing or orbital alignments, but the structure. Heavy metals are drawn inward. Lighter gases were flung to the edges. Planetary bodies falling into zones as if following cues. The Theory of Everything offers a deeper logic: gravitational resonance directs the process from the beginning.

Gravitational Resonance – The Universe's Tuning Fork

Imagine a tuning fork struck in silence. Sand on a plate begins to shift, forming patterns with no hand to guide them. It's a quiet choreography, a shaping influence that doesn't push, only reveals what's already written into the space.

TOE suggests the early solar system worked like that. Not chaos. Not collision-based gambling. But order. Subtle gravitational waves rippling through the disc, structuring it. These waves weren't random. They selected. They sorted.

Dense elements, attuned to the frequency, spiralled inward. Iron. Nickel. The stuff of rocky planets. Lighter gases drifted outward, pulled into deeper gravitational wells to form the giants.

This removes the need for patchwork explanations. No solar wind excuses. No late-stage collisions to explain the inner and outer planetary divide. Resonance did the sorting early, before the wreckage.

Saturn's Rings – A Living Example

For a glimpse of this in action, look to Saturn. Its rings aren't evenly spread. Some regions are dense. Others empty. Gravitational harmonics determine where material can stay. Resonant orbits allow structure. The rest is pushed out.

Now scale that up. Each forming planet becomes its own resonant node. Frequency determines where it forms and what it becomes.

Mercury, Venus, Earth, and Mars: dense, compact, structured. Jupiter and Saturn: vast, gaseous, gravitational anchors. Each is a product of pattern, not chance.

We see this logic beyond planets. In asteroid belts. In the Kirkwood gaps. In exoplanetary systems that uncannily echo our own. The pattern holds.

Moons and Metronomes – The Clockwork of Jupiter

Jupiter's moons demonstrate this beautifully. Io, Europa, and Ganymede are locked in Laplace resonance. For every four orbits of Io, Europa completes two, and Ganymede one. A gravitational rhythm, not a coincidence.

It is not just neat. It keeps Io volcanic. Europa is warm beneath its crust. The resonance is doing work.

Saturn's moons fall into familiar patterns. Pluto and Charon do, too. These aren't anomalies; they echo the same structural logic written into the system.

Beyond the Solar System – A Harmonic Universe

If resonance shaped this system, where else might it reach?

TOE suggests the principle extends far beyond planets. Spiral galaxies. Filamentary structures stretching through cosmic voids. Perhaps these too are shaped by gravitational harmonics, scaled up across the universe.

Classical models view planetary formation as a local and chaotic process. The TOE suggests a wider architecture: mass, motion, and time interacting to yield form. Structure isn't imposed; it emerges.

The Music of the Spheres – Revisited

So perhaps this old phrase, the "music of the spheres", wasn't just poetic indulgence. Perhaps it was closer to the truth than realised. The solar system isn't a random scattering of rocks and gas. It's an arrangement. A vibration made visible. A structure built on frequencies too low for the ear but woven into the very nature of space.

Planetary Architecture Around Massive Stars: A TOE Perspective

In the gravitational worldview proposed by the Theory of Everything (TOE), the mass of a star is more than a statistic. It becomes a kind of celestial composer, shaping the harmonic architecture of the planetary system around it. A more massive star doesn't just pull harder: it resonates more deeply, sculpting the orbits and composition of its system much like a tuning fork defines standing waves.

Exoplanet data increasingly reflects this: massive stars often host large gas giants that orbit unusually close. The architecture seems intentional, not random. Through the TOE lens, gravity acts less like a brute force and more like a structuring field: a resonant blueprint.

1. Fewer, Larger Gas Giants

Massive stars tend to produce fewer, but more dominant, gas giants. This is not likely a sampling error. It suggests that resonance doesn't simply gather matter; it filters and focuses it. Possibilities condense into singular, oversized outcomes. These giants are not just large; they define the system.

Around smaller stars, planetary formation scatters. Debris and dust form a democratic spread of orbiting bodies. But around massive stars, resonance simplifies. It sculpts. A single dominant gas giant often emerges, occupying space with authority. Planetary formation becomes less like a crowd and more like a coronation.

2. Inner Rocky Planets and Stratified Order

In these systems, inner rocky planets often follow a coherent pattern. Heavy elements spiral inward, not randomly, but in a way that appears orchestrated. Standard models can explain density gradients, but TOE deepens the idea, suggesting matter falls into resonant zones, not just by mass, but by harmonic interval.

These planets resemble notes in a chord. Their orbits are not merely stable, but spaced with symmetry, as if shaped by a gravitational rhythm rather than collisions.

3. Consolidation of Ice Giants

In less massive systems, outer planets like Uranus and Neptune emerge as distinct icy bodies. But around massive stars, this diversity tends to vanish.

Resonance, in the TOE model, reduces complexity. It pushes toward simplicity. Instead of multiple ice giants, there may be one super-Neptune or none at all. Outer material either consolidates or is absorbed into the growing gas giants. It's not disorder; it's gravitational economy. A system that favours cohesion over excess.

4. Resonant Rings and Moon Systems

Moons are usually dismissed as planetary sidekicks. But in the TOE view, they hold their own: harmonic companions formed in step with their worlds, carved by the same unseen contours of space.

Jupiter's Galilean moons. Saturn's ring divisions. These aren't random outcomes. They show patterned motion, locked into orbital ratios that suggest underlying structure. Around massive stars, these patterns may become even more pronounced. Moons wouldn't simply orbit. They'd fall into patterns shaped by the deeper geometry of the system itself. Only certain configurations hold

5. Empirical Echoes in the Data

Surveys like Kepler and TESS continue to show that large stars often host hot Jupiters, giant planets in close orbits that challenge standard models. Typically, this is explained through migration. But what if these planets didn't migrate?

TOE proposes that they formed precisely where resonance permitted. Not pushed inward by chance, but pulled into position by gravitational harmony. The data doesn't defy theory; it begins to validate a deeper one.

In this view, planetary architecture isn't an accident of accretion. It's structure, quietly encoded in space, an underlying logic mistaken for disorder.

Smaller Stars and the Optimal Conditions for Life (TOE)

The Theory of Everything doesn't favour spectacle. It suggests that the best conditions for life may not arise around massive, radiant stars, but around their quieter, dimmer cousins. Stars slightly smaller than our Sun. Many of them red dwarfs. Could offer the most reliable foundation for complexity. Not despite their subtlety, but because of it.

1. Planetary Stability and Composition

Around smaller stars, the gravitational pull is gentler. Planetary systems form differently. Rocky planets aren't displaced by gas giants or flung into unstable orbits. They find room to settle.

There's a kind of architectural restraint here. With less material being drawn in aggressively, the formation becomes calmer. Less turbulence. More cohesion. Solid surfaces and stable atmospheres. Conditions typically associated with habitability are more likely to emerge and persist.

2. The Gift of Time

Red dwarfs live on staggering timescales. Tens of billions, even trillions of years. These numbers stretch the imagination, but their implications are grounded: if life arises, it has time.

Larger stars burn fast and die violently, often erasing what they once sustained. Smaller stars burn slowly, granting evolution the canvas of time. TOE sees this not as luck, but as a pattern. A gravitational cadence favouring depth over spectacle

3. Habitable Zones: Closer, Safer

The habitable zone around these stars lies much closer in. Initially, this seems problematic; risks of tidal locking or flares. But TOE offers another view. If the gravitational rhythm is right, proximity becomes an asset.

Resonance in such systems is subdued. Gravitational forces are steadier. Fewer spikes. Fewer disruptions. Conditions can remain stable for billions of years, providing the continuity that life needs to take root and evolve.

4. Minimal Gas Giant Disruption

In large systems, gas giants often dominate. They scatter smaller planets, redirect debris, and disrupt orbiting stability. Around smaller stars, gas giants are rarer and less influential.

This shift matters. Gravitational balance, in TOE's framework, is more than chance. It's essential. Without a Jupiter-like presence to destabilise things, smaller planets can remain undisturbed, orbiting quietly, evolving slowly.

A Final Thought

So, where is life most likely to emerge? According to TOE, not around stars that dazzle. But around those that endure quietly. Their subtle gravitational signature creates space for complexity to arise.

And current observations support this. We're detecting rocky worlds in close, calm orbits around long-lived stars. They're stable. They're ancient. And they're common.

13. Rings Around Young Stars, or Protoplanetary Disks (TOE)

There is something quietly striking about the rings we see around young stars. Not just because they're beautiful, but because they seem too structured to be random. These aren't loose clouds of gas and dust. They are layered. Organised. And if the Theory of Everything is correct, they reveal the early influence of gravitational resonance shaping matter before planets exist.

Protoplanetary disks are vast, rotating bands of material encircling young stars. Observatories like ALMA have revealed rings with unexpected clarity: sharp edges, evenly spaced voids, and patterns that look less like chaos and more like a blueprint.

Take HL Tauri, TW Hydrae, and PDS 70. Each hosts a disk that isn't smeared or shapeless. Instead, we see clean concentric bands, segmented and separated by precise-looking gaps. Standard models might explain this using hydrodynamic instability or the first signs of planet formation. Those explanations have merit, but TOE adds another dimension.

It proposes that these rings are a physical expression of resonance. Gravitational wave harmonics, moving through space, shape the distribution of mass. Matter begins to collect at nodal points, like ripples locked into place on a drumhead. Only here, the waves are four-dimensional.

Resonance

In this framework, gravity doesn't just pull, it sorts. Denser materials settle closer in, while lighter particles drift outward. Over time, this stratification enables the emergence of complex planetary structures, including metallic cores, rocky mantles, and icy exteriors. Saturn's rings may offer a scaled-down glimpse of this same principle: layered, patterned, self-organising. Planetary formation, then, isn't a game of chaotic impacts. It's a kind of cosmic filtration, where matter collects not by chance, but by rhythm.

A Universal Template?

If true, this process should be widespread. Not a local curiosity, but a galactic constant. That could explain why so many planetary systems resemble each other. Rocky worlds near the star. Gas giants further out. Symmetrical distributions that appear again and again.

Rather than randomness, what we may be witnessing is gravitational resonance creating repeatable structure. The same cosmic rules are playing out on different stages.

A Brief Catalogue of Evidence

1. HL Tauri: Clear ring structures, with distinct gaps that hint at the gravitational presence of forming planets.

2. TW Hydrae: Similar disk patterns, also featuring interruptions that suggest ordered layering.

3. PDS 70: Particularly compelling. Gaps in the disk align directly with visible, developing planets.

Closing Thoughts

What are these rings, really? Are they clouds, pulled into formation by mechanical chance? Or are they the visible traces of gravity expressing itself, an organising principle shaping planetary systems from the start?

TOE supports the latter. That which we see in these disks is not randomness, but resonance expressing itself. A subtle but persistent order. These aren't just protoplanetary disks. They are scaffolds. Blueprints. The quiet beginnings of structure, and maybe, in time, of life.

Chapter 17

The Theory of Everything VI

A Unified Model for Planetary Formation

There is a long-standing idea that planetary systems emerge from chaos: violent collisions, molten debris, a storm of randomness that eventually calms into something structured. But what if it is not chaos that shapes solar systems, but order? The Theory of Everything suggests that gravity, especially in the form of gravitational resonance, is not a passive background force. It is the quiet architect of planetary formation.

1. Gravity as Organiser, Not Afterthought

Traditional models describe the early solar system as chaotic: planetesimals constantly colliding, merging, or being ejected. TOE offers a different view. A structured unfolding. Material in the protoplanetary disk behaves according to resonant principles, not randomness. It self-sorts by gravitational rhythm, guided by distance from the star and the collective mass of surrounding matter.

The result is a layered system. Rocky planets close to the star. Gas giants in mid-range orbits. Icy bodies further out. This pattern appears not just in our solar system, but throughout observed planetary systems. It suggests that a deeper structure is embedded in spacetime itself.

2. Rethinking the Protoplanetary Disk

Standard explanations lean on temperature gradients: rocky material survives near the star, ices and gases further out. But TOE suggests that gravitational resonance alone can create this distribution. No need to rely solely on heat thresholds or pressure boundaries. Instead, the gravitational field creates resonant zones where matter collects, stabilises, and begins to build planets.

It simplifies the story, reducing the number of variables required to explain planetary architecture. And as exoplanetary data accumulates, the resonance model appears to hold more weight.

3. Built-In Stability

Planetary systems are often thought to be fragile, their long-term stability treated as a lucky accident. TOE challenges that. It argues that systems form with inherent stability. The star's gravitational rhythm acts like a metronome, aligning orbits into coherent patterns. Planets do not survive by avoiding chaos; they form through its absence.

Different star types seem to produce different system layouts. Large stars tend to host fewer, more dominant planets. Smaller stars often feature more compact rocky systems. The pattern is not absolute, but it shows a clear bias toward order.

4. A Tool for Prediction

If resonance governs system formation, it becomes a predictive tool. With knowledge of a star's mass and age, we can begin estimating the likely arrangement of its planets. Observational data increasingly supports this. Many systems reveal spacing and size distributions that match resonance-based expectations, not perfectly, but often enough to raise eyebrows.

The planetary orbits in some systems even resemble musical intervals, as if the star's gravity has drawn planets into harmonic alignment.

5. A Less Violent Cosmos

The broader view TOE offers is a philosophical one. It suggests a universe where structure arises from order, not from chaos. Complexity doesn't emerge from destruction, but from resonance. This doesn't negate entropy or instability, but it reframes planetary systems as products of coherence. Solar systems may not be precarious accidents. They might be natural outcomes of a universe tuned to create rhythm, pattern, and longevity.

TOE's Potential to Simplify Orbital Dynamics

In physics, complexity is often mistaken for truth. Theories become dense, cluttered with equations and corrections, and we assume the messiness reflects the complexity of the cosmos. But what if clarity has been hiding in simplicity all along? The Theory of Everything (TOE) suggests just that: a gravitational resonance, subtle but universal, might underlie orbital behaviour from electrons to galaxies. If correct, orbital dynamics, once treated as chaotic and computation-heavy, could be reframed as the result of a single, coherent principle.

Why TOE Could Simplify Orbital Dynamics

1. Uniformity Across Scales TOE proposes that gravitational resonance operates at all scales. Whether it's subatomic spin or planetary revolution, the same pattern may be at work. This removes the need for separate models across domains. Rather than switching tools for the quantum and the cosmic, we may be seeing variations of a single law, expressed in different forms.

2. Elegant Mathematics

Existing orbital models function, but with caveats. They rely on approximations, perturbations, and constant adjustment. TOE implies that once gravitational resonance is fully understood, a simpler, more refined equation may replace this patchwork. Not more mathematics, just better mathematics, fewer moving parts, fewer post-hoc corrections.

3. Predicting Orbits

More Easily Today's simulations of multi-body systems demand massive computational resources. The three-body problem alone resists clean solutions. TOE offers another route: systems might not need brute-force calculation if resonance guides them into natural equilibrium. Celestial bodies could self-organise, falling into long-term orbits governed by resonance, not randomness.

4. A Unified Framework

Astrophysics is full of specialised models: one for satellite motion, another for galactic spirals. TOE consolidates them. If gravitational resonance is foundational, then patchwork solutions like dark matter may be placeholders, not necessities. From planetary rings to galaxy rotation curves, the same principle could be quietly at work throughout.

5. Intuition Reconnected with Theory

Physics isn't just about precision. It's about insight. TOE doesn't merely predict behaviour; it explains it. In this model, orbital systems aren't precarious or lucky ,they're structured by resonance, which creates stability by design, not coincidence. Celestial mechanics begins to feel intelligible, not just computable.

Implications for Orbital Mechanics

1. Self-Stabilising Systems

TOE suggests that resonance creates internal order. Planets settle into stable orbits because the system organises itself, not because each force is perfectly balanced in a chaotic dance.

2. Multi-Body Regularity

Systems that appear unstable, like triple stars or galactic interactions, might reveal hidden coherence. Resonance may underlie what now appears chaotic, forming patterns beyond current gravitational models.

3. Streamlined Equations

The complexity of orbital dynamics might be reduced to simpler forms. Rather than endless calculations, a clearer understanding of resonance could make predictions faster, more accurate, and less reliant on computational brute force.

4. Unification Across Scales

This isn't just about planets. TOE proposes that the arcs of comets, the structure of galaxies, even the behaviour of particles, could be governed by the same gravitational phenomenon..

Additional Insights

1. Simplicity Often Hides Truth

Science progresses when chaos gives way to clarity. TOE continues this tradition, asking whether the universe has always been ordered by something simple, something elegant, just hidden behind our models.

2. Current Models Are Provisional

While modern orbital theories are functional, they remain approximations. TOE reframes them as the first draft, not the final word. The real insight may come not from refining corrections, but by uncovering the principle that makes corrections unnecessary.

3. A New Take on the Three-Body Problem

Rather than endless unpredictability, TOE implies a rhythm beneath the noise. Resonance may act as a stabiliser, allowing multi-body systems to persist in patterns we've yet to mathematically resolve.

4. Gravity as a Feedback System

In this model, gravity doesn't just pull; it adapts. It guides systems toward balance, self-correcting when instability arises. It becomes a dynamic organiser, not just an attractive force.

5. Balancing Mathematics with Meaning

The most profound theories offer both precision and understanding. TOE refines our tools, but also restores a kind of human clarity. It suggests that gravity may not just be a computation; it communicates, in a language simple enough to grasp.

Conclusion

TOE doesn't just improve orbital mechanics. It reframes the entire field. If gravitational resonance is real, then our models are not wrong; they are simply incomplete. We are not refining Newton or adjusting Einstein. We are discovering something more profound: that the universe prefers harmony. Orbits, in this view, do not occur by chance.

And perhaps the most significant revelation is not that this makes our calculations easier, but that it makes the cosmos intelligible; a resonant, self-organising order that doesn't just move, but makes sense.

Accretion Discs: Resonance at the Threshold

Near the event horizons of black holes, matter spirals inward not in quiet surrender but in a storm of energy and compression. In these glowing accretion discs, dust and gas are caught in a final orbit, heated to extremes by friction, compacted by gravity into radiative collapse. Conventional physics treats this mostly as a thermal process

: pressure, viscosity, centrifugal balance. But perhaps something deeper pulses beneath the noise.

What if this chaos is not disorder, but resonance? A gravitational resonance so intense that it compresses quantum space itself, distorting the fabric of spacetime into a trembling superposition. At these scales, where quantum fields and relativistic curvature collide, space may no longer remain passive. It could shudder like a drawn string under tension, vibrating with inward gravitational waves. In this light, accretion discs become more than sites of energy loss. They are expressions of an emergent symmetry, a self-organising turbulence driven by resonance.

Turbulence, then, is not a breakdown of order but its frontier - a boundary where gravitational resonance amplifies quantum fluctuations into radiative architecture. The disc becomes a liminal structure, where classical rules begin to falter and quantum geometry asserts itself. The edge of chaos is also the edge of coherence.

Neutron Stars: Rhythm from Collapse

Neutron stars sit like the punctuation of stellar life - the compressed remnants of giants, stripped to their cores, where atoms collapse and electrons fuse into neutrons under crushing gravity. But it's not just their density that holds attention. It's their rhythm.

Pulsars, their spinning counterparts, flash beams of radiation with clock-like consistency. Some are so regular we use them to mark time. Conventional physics attributes this to angular momentum and magnetic radiation. That fits on paper, but it leaves a question hanging. What locks the rhythm in place so precisely? What resists the drift?

If gravitational resonance shapes atoms and orbits, perhaps it holds these stars in tune as well. Perhaps the spin of a pulsar is more than a mechanical echo - it's resonance made visible. Quantum threading through collapsed matter, maintaining order where disorder might otherwise reign.

These are not just dead stars. They are instruments playing the deeper music of spacetime. Not noise, but cadence. Not chaos, but an enduring pulse. And in the silence between rotations, there may be more structure waiting to be heard.

White Dwarfs: The Quiet Geometry of Cooling

A white dwarf is a star at the end of its life. No more fusion. No dramatic collapse. Just a slow fading, a release of residual heat that takes billions of years. The textbooks call it thermal decay, entropy playing out in silence.

But even our own Sun defies thermal simplicity. Its corona burns hotter than its surface, a fact that standard models still struggle to explain. Magnetic fields, plasma loops, these are the usual suspects. Yet something deeper might be at play.

Gravitational resonance, if it shapes the geometry of space, may still linger in the dwarf's heart. Even in apparent stillness, resonance could guide the way heat escapes, subtly directing how energy moves through a star no longer burning.

This is not just a corpse cooling in the dark. It may be an object still in conversation with the universe, still humming at a frequency just below

Magnetars: Tension at the Limits of Matter

Magnetars defy normal categories. Take a neutron star, already extreme, then twist its magnetic field to absurd intensity - a million billion times Earth's. The result is an object so volatile it can knock satellites offline from across the galaxy. It doesn't just radiate; it lashes out.

Standard theories invoke magnetic reconnection or crustal stress. But these feel like surface stories. A magnetar feels different. It behaves like something under strain, something wound too tight.

If gravitational resonance threads through all scales, what happens when it becomes trapped in such a dense, twisted form? What if the magnetar is not just a magnetic monster, but a resonant crucible? A place where the geometry of spacetime is under such pressure that it begins to distort, to pulse, to fail and recover in irregular cycles.

These flares may not be breakdowns. They might be attempts at rebalancing. Outbursts from a body trying to tune itself back into coherence. Not anomalies, but signals from the deeper architecture of the universe. These stars are not misfits. They might be keys.

Proposed Tests – Experimental Validation of Gravitational Resonance

Why Traditional Cavendish Experiments May Be Limited

The Cavendish experiment, long hailed for measuring gravity, rests on a simple premise: that gravity is a direct force between masses. But what if that's only the shadow of a deeper structure?

In the TOE framework, gravity isn't a classical force. It's the result of inward resonance from a higher-dimensional field. Matter isn't pulled; it's positioned. Stabilised. The attraction we record may be a by-product, not the source.

Cavendish assumed a closed mechanical system: mass attracts mass. But if resonance is environmentally or quantum-sensitive, what he measured was only a fragment: a whisper, not the full harmonic.

It's like analysing a piano note with a tuning fork. Useful, yes, but deaf to the resonance of the entire instrument.

1. Gravitational Wave Detection Using Bose–Einstein Condensates

This test isn't about brute measurement. It's about listening more closely. If gravitational resonance exists as a high-frequency, inward-moving wave, it may reveal itself through subtle interactions with sensitive quantum systems.

Objective: To detect high-frequency gravitational waves (above 1 MHz) by observing their influence on sound propagation within a Bose–Einstein condensate (BEC). The goal is to capture resonance shifts at the quantum level; shifts that might evade conventional detectors.

Materials and Setup

1. Condensate Preparation A stable BEC made from rubidium or sodium atoms, cooled to near absolute zero using laser and magnetic trapping. These atoms are ideal for their stable, well-understood quantum states.

2. Resonance Calibration

The BEC's natural vibrational modes are mapped in detail, maximising the system's sensitivity. Precision is critical; too much noise, and any signal vanishes into uncertainty.

3. Gravitational Wave Simulation

Controlled perturbations are introduced via laser interferometry or synthetic resonance generators, spanning a range of frequencies to probe different windows of interaction.

4. Detection and Observation

Sensors are tuned to measure oscillatory shifts, amplitude, frequency, or phase within the condensate. Any coherent shift aligned with an input signal would indicate possible gravitational resonance coupling.

5. Data Cross-Referencing

Patterns are compared against input signals to rule out randomness. The aim is not isolated anomalies but consistent, repeatable alignments.

6. Verification Through Repetition

Trials are conducted under varied conditions to confirm robustness. Reproducibility becomes key; coherent behaviour across trials would elevate the finding from anomaly to phenomenon.

Expected Outcome

If a consistent correlation emerges between gravitational input and resonance shifts in the BEC, it would mark a turning point. These waves wouldn't be passive background noise; they'd be active participants in quantum behaviour. Gravity and coherence, long assumed to inhabit separate domains, may be two aspects of the same underlying structure.

2. Time Crystal Experiments in Microgravity

Time crystals, a strange class of quantum systems that exhibit perpetual oscillation without external energy input, offer a compelling opportunity to investigate gravitational resonance. But what happens when these systems are removed from Earth's gravitational field and placed into microgravity? Could their behaviour reveal deeper insights into the link between gravity and quantum coherence?

Running such experiments aboard the International Space Station may offer unique advantages. Without Earth's gravitational baseline, time crystals might interact more directly with subtler aspects of spacetime, potentially revealing sensitivity to gravitational resonance. If these systems display altered oscillatory patterns or shifts in coherence, it could suggest that gravity plays a more intricate role in quantum stability than previously thought.

This would imply that resonance not only organises stars and planets but may also help sustain the coherence of quantum systems. Any change in behaviour triggered by microgravity or gravitational fluctuation might indicate that gravity is not merely an external force but a stabilising framework embedded within quantum matter.

If confirmed, this sensitivity could also deepen our understanding of biological coherence. It may suggest that the persistence of life and structural order, from atoms to organisms, is tied to the same gravitational resonance that shapes the cosmos.

3. Biological Tests in Microgravity: Photosynthesis and Spiral Patterns

Life has always moved in step with the quiet influence of gravity. But what happens when that influence fades? Could gravity be more than a force, a resonant field shaping biological function at the quantum level?

This proposal explores that question directly. By observing photosynthesis and spiral pattern formation in microgravity, we aim to reveal gravity not as a constraint but as a guiding influence on life's architecture. The International Space Station, where gravity is reduced but not absent, offers a rare setting to test whether life responds to subtle shifts in the shape of space itself.

Objectives

1. To assess whether gravitational resonance affects the quantum mechanics behind photosynthesis.

2. To investigate whether gravity influences the spontaneous emergence of spiral patterns in living systems.

Photosynthesis Experiment

1. Plant Selection

Test organisms will include Arabidopsis thaliana, a well-studied model plant, and Chlorella vulgaris, a photosynthetically efficient alga. Both have demonstrated viability in space and are sensitive to subtle environmental shifts.

2. Gravity Modulation

Two growing conditions will be used. One group will develop in microgravity aboard the ISS. The other will grow in a centrifugal chamber simulating Earth gravity. Additional settings will mimic lunar and Martian gravities to provide a gradient of conditions.

3. Environmental Controls All other variables: light spectrum, CO_2 levels, temperature, and nutrients, will remain constant. Only gravity will vary.

4. Monitoring Quantum Photosynthetic

Activity Measurements will include chlorophyll fluorescence to assess Photosystem II efficiency, gas exchange to track O_2 and CO_2 flux, and electron transport rates to detect quantum transitions during light-energy conversion.

5. Hypothesis

If gravitational resonance plays a role, then measurable deviations should appear in energy conversion rates, excitation efficiency, or redox behaviour under altered gravity. These would suggest a deeper gravitational link to the quantum processes underlying photosynthesis.

Spiral Formation Experiment

1. Biological Models

From coiling algae to root tip spirals and mollusc shells, life often follows helical paths. This experiment will test whether such structure is gravity-dependent or resonance-driven.

2. Experimental Conditions

Both microgravity and centrifugal environments will be used. In microgravity, the absence of standard gravitational cues may disturb spiral formation. In the centrifuge, inertial acceleration will test whether it can replicate gravity's organisational role.

3. Morphological Assessment

Spiral symmetry, coiling radius, chirality, and growth trajectory will be analysed using high-resolution imaging and fractal modelling.

4. Hypothesis

If gravity acts as a resonant scaffold, spirals may become irregular or fail entirely in microgravity. If rotational force does not restore their form, it would suggest that gravitational mass and inertial motion are not interchangeable in biological systems.

Expected Outcomes

1. Photosynthesis Reduced efficiency or shifts in energy conversion under microgravity could expose gravity's role in quantum biological activity. Such changes would challenge traditional views and suggest gravity is involved at far subtler scales than once thought.

2. Spiral Morphogenesis

Disrupted spiral development would support the idea that resonance, not just force, guides pattern formation. If centrifuge-induced gravity cannot fully restore these forms, it may confirm that gravity's effect on life is more than mechanical.

Conclusion

These are not typical biology experiments. They press into a frontier where gravity, quantum coherence, and life intersect. If photosynthesis and spiral morphogenesis show sensitivity to gravitational resonance, the implications would be wide-reaching.

It would suggest that life does not simply adapt to gravity, but is shaped through it, patterned by a resonance woven into the geometry of space itself. Life would not exist solely within the universe, but would reflect its deeper structure.

4. Gravitational Resonance and Its Potential Biological Effects

Before proposing new tests, it's worth noting that ***indirect confirmation*** of the TOE framework may already exist.

In NASA Technical Paper *Gravitational Biology: The Rat Model* (19980024367), data from the Bion-3 and Bion-4 spacecraft revealed something quietly radical. Rats exposed to artificial gravity through onboard centrifuges retained more bone density than those in microgravity—yes, but still showed measurable degradation compared to Earth-based controls.

What matters.

This wasn't zero gravity versus 1G. It was simulated 1G. Centrifugal force engineered to replicate Earth's gravitational pull. And yet... these weren't fragile test subjects. They were the Spartans of the lab world; biological tanks, bred for resilience. Still, they cracked. Not dramatically. But consistently. And exactly where TOE would expect them to.

What did they show?

1. Bone density loss
2. Muscle atrophy
3. Deviations from physiological baselines
4. Partial, not complete protection

In other words, inertial simulation helped, but didn't restore coherence. The rats didn't break entirely. But they degraded. Artificial gravity, it turns out, isn't gravity. Not in full.

It's also worth noting that in the Bion missions, the rats were not continuously subjected to artificial gravity. They were restrained within the centrifuge chamber for only part of the mission, and likely underwent routine exercise protocols; yet, degradation still occurred. This implies that even *with* periodic inertial loading and physical activity, biological coherence was not maintained.

Proposed Resonance Test Framework

1. Initial Calibration – Sea-Level Baseline

A terrestrial study sets the control. A resonating device, emitting low-frequency vibrations or sound waves, is applied to human or mammalian subjects over a fixed period. Physiological metrics are recorded, focusing on muscle mass, bone density, wound healing, neurological response, and cognitive performance. In parallel, plants are monitored for changes in growth and cellular organisation under the same resonance exposure.

2. Space Station Experiment

Astronauts continue using the resonating device while in orbit. The reduced gravitational environment allows comparison. Physiological data are gathered, including muscle and skeletal assessments, tissue regeneration, and cognitive testing. Plants grown in microgravity with resonance exposure are evaluated for differences in growth structure, cellular symmetry, and systemic coherence.

3. Control Phase – No Resonance

Following the resonance phase, the same participants undergo an identical period in microgravity without resonance stimulation. The contrast between these two phases reveals whether resonance mitigates the physiological degradation commonly seen in space.

Hypothesis

Gravitational resonance and the quantum emergence it may foster could be central to biology. In the reduced gravity of orbit, where physical load-bearing is minimal, resonance may act as a stabilising influence: preserving muscle mass, aiding tissue repair, and maintaining neurological function. If correct, this would imply that life is not merely adapted to gravity, but structurally entangled with the resonant nature of spacetime itself.

Supporting Evidence

1. Neurological Function

Resonance is known to influence brain activity. Music therapy has shown efficacy in treating neurological disorders such as Alzheimer's and Parkinson's. Focused ultrasound is being trialled for mood disorders and neural repair. These therapies rely not on chemical modulation, but on vibrational energy, suggesting that the nervous system responds directly to resonance.

2. Healing and Regeneration

Low-frequency electrical and acoustic stimulation has been shown to enhance healing. Frequencies between 100 Hz and 1000 Hz accelerate chronic wound repair. Low-intensity pulsed ultrasound is commonly used to stimulate bone regeneration, suggesting that biological structures respond favourably to finely tuned vibrational inputs.

3. Resonant and Quantum Signatures in Biology

Cells and molecules exhibit natural oscillatory patterns. These internal rhythms may represent a form of quantum tuning, allowing organisms to remain within energetic ranges optimal for life. Phenomena such as stochastic resonance in neurons suggest that weak external signals--including gravitational resonance--can improve biological signal detection, refining perception and coordination.

4. Biological Response to Resonance

Plant biology offers compelling evidence that acoustic and vibrational inputs modulate growth. Some plants increase nectar production when exposed to frequencies mimicking pollinators. Others adjust root growth, cellular symmetry, and fungal symbiosis in response to low-frequency stimulation. These effects persist even under altered gravity, suggesting that resonance, not mass alone, plays a central role in biological patterning and development.

5. Atomic Clocks in Variable Gravitational Conditions

Atomic clocks, the most precise instruments ever built to measure time, may hold clues to a more profound truth. In this framework, timekeeping is not a neutral mechanical process. It may be subtly shaped by gravitational resonance. The proposal is simple: deploy atomic clocks across varying gravitational environments — sea level, high-altitude observatories, and orbital platforms- and observe their synchronisation.

The standard model accepts that gravity affects time, as shown by general relativity. Clocks tick faster at altitude, where gravity is weaker. But this proposal takes a step further. It suggests that what we call time may be a resonance effect. Not merely a warping of spacetime, but an emergent property of atomic energy levels responding to gravitational interference patterns.

If precision timing drifts in ways not fully accounted for by elevation alone, if resonance frequency shifts begin to correlate with gravitational intensity, then we may be observing more than relativistic time dilation. We may be seeing time as a phenomenon that arises from a quantum-gravitational substrate.

6. Microgravity Tests: Double-Slit and Bell Experiments

1. Double-Slit Experiment in Microgravity

Hypothesis The double-slit experiment shows us that particles behave like waves, collapsing into form only when observed. But what if gravity is quietly participating in this collapse? On Earth, gravitational resonance saturates every interaction. In orbit, where that resonance weakens, what happens to the interference pattern?

Objective To observe whether wave-particle duality behaves differently in microgravity. Does the interference pattern remain stable? Or does it shift, fragment, or collapse more easily? If so, this would suggest gravity is not just a passive backdrop, but a hidden actor in quantum behaviour. It may shape the very conditions under which observation becomes measurement.

2. Bell Test in Microgravity

Hypothesis

Quantum entanglement defies classical intuition. But might gravity, or its resonant influence, help anchor the coherence we observe between entangled particles? If gravitational resonance plays even a minor role in stabilising entanglement, then freeing particles from Earth's field may expose that connection.

Objective

To run a Bell test aboard the ISS or in deep space, and measure whether the correlations between entangled particles remain consistent. A measurable deviation could point to a background structure that has so far gone unmeasured - a quantum-gravitational resonance field that shapes the behaviour of what we currently call non-locality.

Rationale for Testing in Microgravity

These are not hypothetical exercises. They challenge the separation between quantum mechanics and gravity--a division that has shaped physics for over a century. But what if that split is illusory? What if gravitational resonance is the quiet link between the two?

Microgravity provides an acoustic void: a near-silence where Earth's gravitational hum recedes. It is in that stillness that subtle patterns may emerge. If wavefunctions behave differently, if entanglement becomes unstable or shifts, we may need to reconsider the nature of reality itself.

Not as separate layers of law, but as one continuous field, resonating across all scales. From the flicker of atoms to the sweep of galaxies, the same underlying rhythm may be at play; not bending space, but shaping time and form through resonance.

Theory of Everything Model: Calibrating TOE to Explain Celestial Anomalies

A Theory of Everything grounded in gravitational resonance and quantum compression fields invites us to rethink gravity, not as a force transmitted across a vacuum, but as a resonant phenomenon embedded within the very scaffolding of space itself.

Rethinking Planet Nine: A Quantum-Gravitational Perspective

Something curious occurs in the outer reaches of the solar system. Trans-Neptunian Objects drift along orbits that defy conventional expectations. To explain this, astronomers proposed the existence of Planet Nine: a hypothetical massive body tugging unseen from the shadows. But perhaps what's missing is not a planet, but a piece of understanding.

Through the TOE framework, these orbital irregularities may not indicate hidden mass but reveal quantum compression fields, non-local distortions in the energy density of space. These are not exotic particles or placeholders for dark matter. They are shifts in geometry, subtle resonance effects arising from gravity's behaviour in more than three dimensions. They wouldn't show up in a telescope, yet could produce exactly the orbital nudges we observe.

In this view, space is not inert. It responds. It vibrates. It shapes matter not through force alone, but through structure and resonance. What appear to be gravitational anomalies may be echoes of a deeper pattern in the fabric of space.

Solar System to Galactic Rotation: A Challenge to Dark Matter

On a larger scale, the mystery deepens. Galaxies rotate too quickly, their stars whirling as if immersed in an invisible halo of mass. Dark matter has long served as the explanation. But what if the anomaly lies not in what's missing, but in what we assume?

TOE proposes that quantum compression fields extend across entire galaxies. These fields distribute gravitational resonance outward from galactic cores, mimicking the effects of additional mass. Gravity, in this model, is not merely emitted by matter. It arises from the resonant geometry of space itself. The result is a natural reproduction of galactic rotation curves, no unseen particles required.

Cosmic Expansion: A Function of Quantum Gravitational Resonance

The universe's accelerating expansion, commonly attributed to dark energy, may also be reinterpreted. What if expansion is not driven by an external repulsive force, but by resonance differentials within space itself?

Rather than uniform growth, TOE envisions a cosmos pulsing with variable-density quantum fields. Localised intensities in gravitational resonance could amplify or restrain expansion depending on spatial context. This perspective not only accounts for the acceleration but also explains its variation across time and region.

Predicting Gamma-Ray Bursts and Cosmic Flashes

If gravitational resonance underpins structure at all scales, it should also leave its mark on the most transient cosmic events. Gamma-ray bursts and unexplained flashes may not be chaotic. They could be resonant signatures, the result of gravitational wavefronts intersecting and concentrating energy at nodal points.

Like converging waves in a sea, the pattern is complex, yet not random. These bursts become signals, not anomalies, revealing the undercurrents of cosmic structure.

A Unified Gravitational Model: Beyond Hypotheticals

This model does more than reinterpret observations. If TOE explains TNO orbital shifts without invoking Planet Nine, accounts for galactic motion without dark matter, and reimagines expansion without dark energy, it offers more than refinement. It points to a unified paradigm where gravity is not a force but a resonance, a modulation of the quantum vacuum itself.

In this vision, gravity conducts the architecture of matter and motion through patterns of coherence, not pressure and pull. It sculpts spacetime through vibration, not collision.

Implications for Black Holes and Cosmic Structure

Black holes, often treated as singularities of infinite density, also shift under this lens. A supermassive black hole may not be a gravitational abyss, but a tuning node. Its immense mass may structure surrounding space through resonance rather than collapse alone.

These are not simply the endpoints of stellar collapse. They may be active players in a deeper lattice of gravitational geometry, orchestrating space, energy, and even cosmic evolution itself.

Conclusion

The TOE model doesn't just close gaps. It opens a new frame of reference: where gravitational resonance, quantum fields, all arise from a single foundational principle. No duct tape. No placeholders. Just a coherent, emergent symmetry vibrating through the entire cosmos.

What once appeared fragmented may have always been whole, a continuum, subtle, precise, and profoundly interconnected.

Space Degeneration and the Role of Gravitational Resonance in Biological Systems

Astronauts aboard the ISS still experience about 89 per cent of Earth's gravity. Yet even this slight reduction leads to serious physiological decline: visual impairment, cognitive deficits, bone loss, muscle atrophy, and accelerated ageing. Traditional explanations point to a lack of mechanical loading. But the severity suggests something deeper is at play.

Why do sedentary patients on Earth fare better than astronauts in microgravity? The answer may lie not in the force of gravity, but in its resonance.

From a Theory of Everything perspective, gravity is not just a downward pull. It is a resonant field, a stabilising hum that maintains coherence in biological systems. Evolution took place within this-low-frequency field. Remove or dilute it, and cellular, neurological, and genomic stability begins to unravel.

Key Observations

1. Neuro-Ocular Syndrome: Changes in vision and intracranial pressure may stem from disrupted cerebrospinal fluid regulation, previously stabilised by gravitational resonance.

2. Cognitive Decline: Gravity may help sustain coherence in neural wavefunctions. Without it, thought itself begins to fray.

3. White Matter Loss and Nerve Dysfunction: Signal transfer and tissue integrity degrade in space. Resonance could be the unseen organising layer.

4. Bone and Muscle Atrophy: Mechanical countermeasures fall short. The missing element may be a vibrational one that supports molecular cohesion.

5. Accelerated Ageing and Cancer Risk: Space-induced oxidative stress and gene dysregulation suggest that gravitational resonance may contribute to DNA repair and cellular longevity.

Even an 11 per cent reduction appears to disrupt these systems profoundly. Artificial gravity attempts have failed not due to weak engineering, but perhaps because they recreate only the mechanical illusion of gravity, not its vibrational structure.

Implications for Long-Term Missions

Mars offers just 38 per cent of Earth's gravity. If coherence collapses at 11 per cent loss, the risks of deeper attenuation could be biologically catastrophic. The problem is not force; it is fidelity. Without gravity's resonant signature, the quantum architecture of life may falter.

Conclusion

Gravity, in this model, is not merely structural. It is informational. It organises biological systems from the quantum level upward. Space travel, without accounting for this, becomes a slow form of biological unravelling.

Optimal Genetic Code: Life as a Universal Constant

The Hypothesis of a Universal Genetic Template

Imagine, for a moment, that life is not a serendipitous anomaly, some molecular accident adrift in a sea of inert matter, but an intrinsic feature of the cosmos itself. Suppose the very fabric of spacetime, those silent gravitational resonances that ripple endlessly through the void, encodes not only geometry and motion, but a blueprint for biology. In such a scenario, life is not a localised miracle but a universal inevitability. Evolution becomes less a chaotic struggle for survival and more an unfolding, an expression, of an embedded cosmic architecture, revealed wherever the conditions are fertile enough to permit it.

The Paradox of Mutability and Constancy

Biological life, paradoxically, is sustained by mutation. It adapts, diversifies, and branches wildly. And yet, the genetic code, the deep language of life, has remained essentially invariant across billions of years. Why constancy amidst perpetual flux?

There are several possibilities, none mutually exclusive:

1. Preordained Efficiency

The genetic code may represent an exquisitely efficient system, one so finely tuned that any deviation would unravel core biochemical processes, collapsing the scaffolding on which life depends.

2. Evolutionary Lock-In

Once a viable code is established, its very ubiquity becomes self-reinforcing. Any organism diverging from this universal grammar risks biochemical incoherence and functional death. Cosmic Resonance: Perhaps this stability isn't just a biochemical artefact, but a manifestation of universal structure, a property encoded in the gravitational harmonics of spacetime, anchoring life to a primordial resonant template. Superficial traits mutate and morph. But beneath the flux, the code persists, silent, ancient, and unmoved; a genetic Rosetta Stone etched into the quantum marrow of the cosmos.

Resonant Geometry and Cellular Organisation

Is it possible that cellular structure itself, beyond mere chemistry, is shaped by cosmic resonance? Consider the elegant geometry of biological forms: the radial symmetry of diatoms, the fractal dendrites of neurons, the delicate architecture of radiolarians. These are not haphazard configurations. They seem almost inevitable.

Choanoflagellates, the microscopic progenitors of multicellular life, exhibit behaviours that hint at a natural leaning toward complexity and cooperation. Might their geometric arrangements reflect not just chemical interactions, but alignment with gravitational or quantum harmonic forces?

Even the pufferfish's enigmatic seabed mandalas, sculpted by wave-like fin movements, feel less like instinct and more like choreography, nature dancing to an unseen quantum rhythm.

The Curious Singularity of Life's Genetic Code

Why, in a universe replete with chemical possibilities, has life converged upon a single genetic system? If alternative biochemistries were viable, wouldn't we expect to find divergent examples, as well as failed or parallel experiments, scattered throughout evolutionary history?

Instead, we find a singular code, conserved, robust, and apparently optimal. This suggests not merely a convenient solution, but a necessary one. The primordial code may have arisen not because it was sufficient, but because it was inevitable, a biochemical basin of attraction shaped by the laws of physics themselves. Any deviant attempt was either non-viable or outcompeted into oblivion. Natural selection, scaled cosmically, may have rendered divergence obsolete.

A Molecular Blueprint Beyond Earth

Is this molecular unity an Earthly coincidence, or does it point toward something deeper? A genetic framework shaped not solely by Darwinian pressure, but by the same fundamental harmonics that govern stellar orbits and atomic spins?

If spacetime is a resonant medium, then DNA may be its biological expression, an emergent pattern dictated by the physics of existence. In this light, life is not the exception. It is the expected outcome of a universe designed, not by intent, but by its own structural coherence, to give rise to complexity.

Complexity from the Outset: A Robust Primordial Code

The standard evolutionary narrative holds that life began as a simple self-replicator, gradually acquiring sophistication. But this view becomes strained when we examine the earliest extant life forms, which already exhibit intricate machinery, ribosomes, error-correcting enzymes, and finely tuned translational systems.

Such complexity implies that life did not build itself up slowly from chaos, but sprang forth with a kind of architectural intent:

1. Optimised from Inception

Early genetic systems display a level of efficiency more suggestive of design than blind chance.

2. Entrenched by Necessity

Once these systems took hold, any deviation would have risked total collapse, cementing their structure across deep time.

3. Preserved by Resonance

The universality of the code may reflect its resonance with deeper physical laws, stability conferred not by molecules alone, but by harmony with the very dynamics of spacetime.

Humanity's Binary Code Experiment: A Computational Analogy

We created binary code, just two digits, and with it, we built a digital civilisation. Despite the availability of richer numerical systems, binary endures for its clarity and efficiency. It is simple, but not simplistic. Powerful, yet robust.

Might DNA be the biological equivalent? An elegant, four-letter code, capable of unbounded complexity. Just as binary allows for infinite computation, so too does DNA allow for boundless biological expression, bounded only by the conditions in which it unfolds.

The Imprint of Purpose in the Universal Fabric

If life's genetic template is not an incidental artefact, but a structural inevitability, then we are confronted by a disquieting possibility: purpose. Not purpose as anthropocentric narrative or divine design, but purpose as embedded tendency, a drive toward organisation, sentience, and self-reflection. The universe may not care, but it does cohere, and in that coherence, life finds a foothold.

This is not mysticism. It is a recognition that physical law, resonance, and pattern might favour life, not by chance, but by construction.

A Cosmos Teeming with Life?

If this hypothesis holds, then life is not the exception. It is the rule. Should we encounter extraterrestrial organisms, they may not be wholly alien but oddly familiar, not solely because of convergent evolution, but because they, too, arise from the same gravitational resonance.

Extraterrestrial Communication and Quantum Resonance

Within the Theory of Everything (TOE), traditional radio transmissions may eventually be viewed as artefacts of a less evolved civilisation: blunt tools fumbling through a cosmos far subtler than we imagined. Vulnerable to cosmic interference, lensing, and interstellar dust, these signals are fragile. More crucially, they're constrained by electromagnetism: linear, slow, and tethered to local mechanics. A truly advanced civilisation, TOE suggests, would bypass these limitations. Its communication would likely exploit gravitational resonance or quantum entanglement; phenomena that transcend classical propagation.

1. Gravitational Resonance as a Universal Medium

TOE contends gravitational resonance isn't a by-product of mass, but a foundational rhythm embedded within the fabric of spacetime. This makes it a candidate for a truly universal language. Mathematics and geometry become more than human abstractions; they're emergent from this resonant substrate. Any intelligent life, anywhere, would encounter the same underlying forms: spirals, ratios, harmonic structures. In this view, communication wouldn't rely on symbolic decoding but on direct recognition: resonance expressed as a pattern.

2. Quantum Communication and 4th-Dimensional

Tunnelling If gravity is a 4th-dimensional oscillatory wave, as TOE suggests, then information needn't crawl across space. It could tunnel through it. Quantum entanglement already demonstrates nonlocal coherence. A sufficiently advanced intelligence might harness this, using resonance states to transmit not across but through spacetime, via higher-dimensional interference patterns.

3. Geometric Signatures in Quantum Fields

If gravitational resonance shapes the quantum vacuum, communication may manifest not as "signals" but as geometric distortions: phase-locked fluctuations, coherent anomalies in quantum noise. These wouldn't be heard, they'd be seen. Perhaps some already have been misclassified, dismissed as noise, the order within overlooked.

4. Detection Technologies for Quantum Signals

To engage with such communication, our tools must evolve. Radio telescopes are blind to TOE's deeper fields. Instead, we'd need instruments that detect quantum perturbations and gravitational modulations, Bose–Einstein condensates, for instance, acting as resonance mirrors. Structured fluctuations might appear as vibrational harmonics: not unlike chords in a song, but composed through the geometry of spacetime.

5. Rethinking SETI: A Quantum Perspective

TOE challenges SETI's core assumptions. If intelligence communicates through gravitational or quantum modulation, we've been listening in the wrong place. We must look not for noise but for coherence: unexpected symmetries in the scaffolding of space. The real challenge is not technological, it's epistemological.

We must rethink what it means to "*receive*" a message. It may arrive not in sound or light, but as a synchrony. A structural anomaly. A ripple that resists entropy.

Conclusion

From the TOE perspective, reliance on radio signals begins to feel oddly provincial. A truly universal language would not be linear, brittle, or bound by distance. It would be resonant: a pattern inscribed in space itself. Not a broadcast, but a modulation. To detect it, we must stop reaching and start tuning. The first message may already be here, not a ping, but a symmetry.

TR

Chapter 18

Constellations of Meaning

Four times through, maybe more. I had gone over the theory again and again. The pages flickered past my eyes, shifting, reforming, demanding something deeper. I had watched countless videos, drawn lines between ideas, traced them in the air like constellations forming and reforming, dissolving into meaning. The theory stretched out, vast, seamless, and somewhere within it, something clicked, something settled into place.

I wrote to TR:

Some aspects of the TOE framework are beyond me, I admit. It slips past the edges of what I know, but even so, I find myself caught in its current. It flows, effortlessly, irresistibly. And the solar system, yes, that part especially, how it came to be, the forces at play, the strange and intricate dance of formation that once baffled me now seems... inevitable. Obvious, even. How odd that what once seemed like chaos now feels so perfectly composed.

I paused, fingers hovering over the keys, then added:

Actually, the entire theory, it's alive in a way, isn't it? The way it links life to galaxies, the grand and the microscopic, mirroring one another in spirals, in movement, in resonance. It's fascinating. And I had this thought, how improbable, how wildly improbable, if not impossible, to construct a theory that ties it all together. Everything. The universe, its form, its breath, its motion, its purpose.

I took a sip of tea, feeling the warmth settle in my chest, before I typed the last lines:

I don't mean to pry, but do you have an equation, something like $E=mc^2$? Some singular expression that contains the whole? How did you piece it all together? It's as if you've seen something the rest of us only glimpse. Obviously, you must have a background in physics. And tell me, are there any physicists who might step forward, who might see what you see?

Love, D.

TR's response arrived swiftly, and as always, it was as grounded as it was thought-provoking:

Good, he wrote. You had an adventure with the theory. I'm confident it holds as a Theory of Everything. It's an Occam's razor outcome, stripped of unnecessary complexity. But of course, I have to remain aware of the possibility of motivational reasoning. There's always the risk of seeing what one wants to see. Ultimately, however, the only real test is how well it stands up to scrutiny. And scrutiny is welcome. This is not presented as a final masterpiece, but as a working sketch, one that, while rough around the edges, I believe captures the underlying framework faithfully enough to endure scrutiny beyond the usual gatekeeping instincts. Those who understand frameworks will recognise the invitation here. Those who seek only finished monuments may miss the foundations entirely.

He continued, addressing my question about his background:

Yes, I have a background in physics, though that was moons ago. Truth was always my thing and the big picture, so I found other disciplines more meaningful, more grounded. Academia, at the end of the day, often felt like a theatre of Comic-Con enthusiasts, grant-jockeys, and well-meaning grifters tangled up in bureaucracy, chasing ghosts, bless 'em. It wasn't a place I was willing to grow old in. Sometimes you have to take the path you're not weighed down on. The TOE wasn't born in the fluorescent corridors of academia; it was shaped in solitude. If it holds, it holds. If it doesn't, it doesn't.

TR added:

A potential equation? Perhaps something like

G_{μν} + R_{μν} = 8π⟨T_{μν}⟩_{quantum} + ΛΨ_{μν}

It's not final, of course, just a rough sketch of where the deeper structure may lead.

I asked how the theory first took shape, and his answer came effortlessly, yet carried the weight of decades of reflection and inquiry:

It began many years ago. I recall asking myself, 'What is the key to the universe?' Or at least a hint. The answer intuitively came: balance. The universe seeks equilibrium in everything, from the formation of galaxies to the stability of planetary systems, all the way down to life itself. But the paradox is, imperfection is part of that balance. Without it, systems stagnate. They cease to evolve. It's the asymmetry, the tension between forces, that keeps everything in motion. Years later, I turned my attention to evolution and natural selection. I wanted to quantify the mechanisms at play, to trace the scaffolding of development itself. But something was missing. Evolution, as we framed it, felt incomplete. Like an orchestra playing without a conductor, intricate, yes, but strangely directionless.

Take savants. Daniel Tammet, Kim Peek. These aren't flukes or neurological mishaps. They're signposts. Evidence of a deeper coherence. Of consciousness resting on a structure far more intricate than natural selection alone can explain. When you observe these individuals, you glimpse something beneath biology, a resonance, a hidden order.

So I asked the question plainly: Why does life consistently evolve? What is the constant? The driving force?

And then it struck me: gravity. Not just as a force but as a resonant architecture, shaping not only the cosmos but also the emergence of complexity itself.

Not the black-and-white Newtonian caricature of it, but something else entirely. A subtler pressure. A sculptor behind the veil. Gravity, in its true form, resonant and dimensional, wasn't just a force; it was a guiding architecture. A hidden symmetry. Once I recognised that, and once I could see it through the lens of the fourth dimension, the entire framework began to unfurl. Patterns emerged where before there was only noise. It wasn't mystical. It was simply real, but obscured, like something obvious you've been overlooking for years, hiding in plain sight.

Also, I'd observed, and, at times, experienced, a particular form of four-dimensional luminosity. Not light in the conventional sense. Not electromagnetic radiation. This was something else entirely. It behaved more like a kind of fluorescent presence, subtle, but unmistakable. It surfaced repeatedly, both in my own experiences and in controlled trials. Either it emerges from the interaction between matter and the fourth dimension, or it emanates from the fourth itself. And it seemed to affect the mind. Which, of course, made sense. The mind is matter, after all.

He described how his understanding didn't emerge from sudden flashes of insight but through methodical refinement, building, testing, and iterating:

I started considering the nature of all forms of 3D light in the electromagnetic spectrum. And I began to see them not as fundamental, but as emergent phenomena, products of deeper, fourth-dimensional light waves. This line of thought naturally led to resonance. How life and nature use resonance not only to transfer information, but to structure and influence matter itself. This was the part that fascinated me most. The idea struck me that resonance, something as fundamental as vibration, could be the organising principle behind both consciousness and the physical universe. Some compelling conclusions began forming about why a 4D light wave might act inwardly on all matter. One source of insight was Einstein. Late in life, he seemed to move beyond the idea of gravity as merely the curvature of spacetime. He hinted at something deeper, something fundamental. He even considered that time itself might be emergent, not intrinsic. It was in reflecting on these ideas that I had a breakthrough: a fourth-dimensional resonant light wave, acting inwardly on all matter. Inert matter, confined to three dimensions, can never fully transcend into the fourth, but it remains within its influence.

Then, something unexpected:

During the trials, I had a few visions that offered strange but useful insights into how gravity may function. As an example, you mentioned Crazy Horse on the forum, his vision of the world appearing soft, almost fluid, while his horse behaved strangely. If gravity acts inwardly on all matter, then perhaps this 'softness' is how the human mind interprets that influence. His horse, being a complex living system, might have exhibited quantum-like behaviour, erratic, unpredictable. A fitting metaphor for what he called *Crazy Horse*. And from there, everything fell into place. The Theory of Everything. How the universe evolves. How life evolves.

Then, without pause, he turned to what lay ahead:

If the theory holds merit? A few physicists come to mind. Roger P, one of the most exceptional minds I've encountered. Logical, but open. Though at this stage, I suspect he'd rather spend his days in reflection than in scientific debates. Another physicist has been quietly orbiting the possibility that the universe might think, a notion that, in its own way, runs parallel to TOE. Different peaks, perhaps, but both gazing at the same sky. There are also others, one or two I have in mind, if all goes well. I'd like to lead the theory's development through an organised effort.

And then, as if shifting from the theoretical to the profoundly practical, he spoke about what comes next:

My focus now is on what lies beyond, understanding dimensions beyond the third and their influence on life. This is where my work is most effective. I have a plan, a nonprofit organisation designed to bring together scientists and individuals of uncompromising integrity, not just to develop the Theory of Everything further but to apply it; its implications extend far beyond physics. The mission isn't just theoretical. It extends to real-world challenges, environmental crises, and humanitarian issues, problems that demand both scientific and ethical clarity. Truth and reason must be at the centre of it all. Because where truth leads, real progress follows. Truth and Love,

TR

It took me a while to digest TR's response, but clarity arrived not long after. His insight and foresight were boundless, and I felt reassured that he had a firm handle on both the Theory and its implications. But there was a question I couldn't shake, the Lost Book of Nostradamus and plate 67 had been flashing in my mind vividly, almost like a revelation. Something about it felt in parallel to the Theory's unfolding, as if it was connected more deeply.

I decided to ask:

TR, one more question: Have you ever come across the Lost Book of Nostradamus, particularly plate 67? It came to me so vividly, almost like a revelation, the Book of Life itself. What are your thoughts on the plate? It feels deeply metaphorical, but I can't shake the sense that it mirrors the Theory in some way.

Love, D

After sending the message, I took a break, stepping out into the garden for a while. The crisp air and the sounds of nature settled my restless mind, and I felt an unmistakable shift in my consciousness, a growing appreciation for the intricate connections in life and the mysteries unravelling into clarity.

When I returned to my laptop, TR had replied:

I'm not surprised you noticed it, D. Yes, I came across it some time ago. Speculation about viewing forward in time is understandable, and, in my view, not entirely baseless. From experience, I'm confident it happens, though rarely.

But, Plate 67 is striking for its symbolic depth. Let me outline what I see based on my framework and knowledge. The crown, for instance, doesn't symbolise literal kingship but rather alignment. The hand on the book is that of a lion, which I interpret as representing natural strength and groundedness, the capacity to hold and comprehend what might metaphorically be called the Book of Life. The long hair, which has been venerated across cultures as a sign of wisdom, signifies the capacity for rational insight, reading the book with clarity and depth.

The wheel conveys a process, gravity, time, and the interplay between them. The gap within the wheel is a divide between dimensions. On the left, we see symbols of the fourth dimension:

duality and the waters of creation, represented by the fish. The ribbons, in turn, depict quantum interaction, the exchange between the fourth and third dimensions. The woman holding the bow, absent an arrow, strikes me as a representation of insight or second sight. The bow is "*will*" under tension. The arrow is awareness.

Her role isn't to act directly, but to perceive with clarity. The colours of her dress, subtle but deliberate, suggest Grace, an intangible influence guiding perception.

The phrase "*One male*". If this Theory, or some alignment it facilitates, brings about a deeper understanding of life and the universe, then its meaning becomes self-explanatory.

As for the other plates, I'll refrain from diving into them now. Suffice it to say, they seem to illustrate an unfolding process, with *Plate 82* marking a new beginning.

This is how I interpret it, D. But I must emphasise the importance of staying grounded. Myths and symbolic systems have shaped human history, sometimes with profound insight, but often at great cost. It's essential that we pivot toward reason and evidence. These are the tools that will truly illuminate our way forward.

Truth and Love, TR

Reading TR's interpretation was like watching constellations align, each insight tracing a luminous thread to the next. The plate, his Theory, and the depth of his perspective seemed to merge effortlessly, a symphony of thought and meaning. I found myself eager to explore further, yet I understood his caution about staying grounded. He was right, mythology and fables, for all their allure, have sown their share of discord and pain throughout history. Science, as TR so rightly pointed out, is the compass we needed to chart our way foward

Smiling, I wrote back:

I must say, I'm completely convinced you've decoded the plate. As we discussed, start sending me your chapters. I'll weave them in from here on.

Love, D

Chapter 19
Self Understanding & Healing

Healing and self-improvement are essential, but the distinction between self-understanding and self-help needs to be made clear, because we are dealing with intelligence and the greater good.

I would argue that intelligent individuals form a minority on this planet. They are not defined by race, gender, or any other surface-level distinction but by depth of thought and clarity of perception. And if we are to counteract decline intellectually, socially, and even evolutionarily, we need more grounded, rational minds, individuals capable of seeing reality as it is, not as they wish it to be.

Self-help offers comfort; self-understanding offers power. One soothes, the other transforms. And the path forward depends on which we choose.

If we are to move forward, we need to step back. A rational starting point would be to prioritise self-understanding and self-healing, not the hollow promises of the self-help industry, but the hard and necessary work of coming to terms with ourselves as we truly are.

The self-help industry, for the most part, is a booming, multi-billion-dollar marketplace, a hustler's paradise that thrives by preying on the vulnerabilities of struggling societies. In places where people's self-worth has been systematically eroded, often to the point of being non-existent, they are left open to what can only be described as passive predation through glossy, saccharine literature. This is an industry that has perfected a blueprint for self-help gurus and their ilk, a manual, in essence, for "how to shake people down and manipulate the masses."

The offer rarely changes: Believe this, and your life will change forever. It's a seductive promise. But more often than not, it's a kind of spiritual junk food, momentarily satisfying, ultimately hollow. Think of the typical self-help pitch as the psychological equivalent of trying to lose weight by ordering a wood-fired gourmet pizza. There's that initial rush, warm, comforting, convincing. For a moment, it feels like progress. But beneath the surface, you're left with more baggage than you started with, emotional, existential, or otherwise. Worse still, many of these works promote a toxic form of positivity, which can quickly descend into escapism. Rarely do they offer practical, sustainable solutions or genuine insight.

True self-development doesn't stem from external consumption. It arises from self-understanding, which often demands the courage to confront uncomfortable truths rather than bury them beneath shallow platitudes. It's found in confrontation. In friction. In the quiet, unglamorous space where you stop performing for yourself.

Right alongside this, there's another multi-billion-dollar industry, driven more by self-interest than genuine desires to help people: the therapist consortium. Truly self-aware and knowledgeable therapists, those wise enough to offer meaningful help, are rare. For the most part, the relief people experience in therapy comes from the simple act of being listened to, or the placebo effect of engaging with what often amounts to psychological pseudoscience.

That's not to say there aren't those who have found real comfort in it, some have. But we can't ignore the bigger picture: a profit-driven culture that feeds off insecurity, turning desperation into dollars. Instead of building people up, it leaves them more disconnected, more uncertain about who they really are, and ultimately erodes humanity's sense of self-worth rather than enriching it.

Seeing through the illusions, the layers of deception, that's one thing. But in a world drowning in despair, what truly matters is finding a way through, not just standing by and watching it happen, but actively seeking resolutions. Over time, I've had to pull myself back from the edge more times than I can count, facing a kind of living death, repeatedly having to salvage my own well-being. And through that, I've found something that works. I don't particularly like calling it a "method" because that makes it sound like just another packaged idea, but sometimes we need structure. So, let's call it the Unlock Method. It's not magic, but it can help to remedy or ease the burden of "nervous exhaustion" disorders such as depression, PTSD, panic, fatigue, and the alleviation of persistent pain, the kind of struggles that are steadily rising in a manufactured reality wearing people down in body, mind, and spirit.

The first step is to understand your own mind, not as some unique, isolated self, but as part of something much older, something shaped by time and survival. You have to step outside of yourself to truly see yourself, or, to put it another way, being impartial towards yourself helps to navigate a pathway to recovery. A good place to start is by watching Carl Sagan's talk on the human brain, "Carl Sagan on Human Brain 360p"-specifically the part about the R-complex, the oldest, most primal part of who we are. That's the foundation. Once you understand that, you can begin to see how the Unlock Method fits into the bigger picture, how it helps untangle the patterns that can become debilitating or undermine self-worth.

It's important to recognise that no method works universally. Individuals are unique, and results may vary. But I want you to know that my approach comes from a place of genuine understanding, because I've lived this, and because I know the power of real self-awareness, and a true desire to help. It may not resonate with everyone, but if you don't try, you don't move. And right now, moving forward in self-understanding and healing is what humanity desperately needs.

I will use a more methodical approach below, because sometimes creativity can get in the way of clarity, and this needs to be straightforward.

The Unlock Method and Healing

To use critical thinking and wisdom to see beyond the veils of duplicitous practices, is one thing. But it's far more rational in a world brimming with despair to offer resolutions. Over the years, I have pushed myself beyond a type of living death, and then needing to salvage myself on many occasions, I have developed a tried and proven strategy.

Method is not a word I like to use, but it's necessary, along with self-understanding. So, we will call it the "Unlock method." It should help remedy a variety of conditions that bring about "*nervous exhaustion*" disorders, such as depression, PTSD, panic disorders, chronic fatigue, easing chronic pain, and so on. A disclaimer: I'm motivated to help you; understand you. The method may not work for everyone, but nothing ventured, nothing gained.

First step is understanding your brain's evolution, this is not about your unique individuality other than to recover it. You need to learn to be impartial to yourself in order to understand yourself.

The Unlock Method: A Guide to Recovery from Nervous Exhaustion

Overview

When the mind and nervous system are traumatised by illness, abuse, or sheer exhaustion, a host of debilitating symptoms can begin to unfold. Much like the body, the mind has its limits. But unlike a broken limb, which can be splinted and left to rest, the mind never really stops. Even in sleep, it hums along. And when the system is strained, sleep can become just another form of waking, a restless theatre of unresolved tension.

You might think of the mind as a camera. When healthy, its lens is clear, and the depth of field is precise. It picks up the subtleties, the nuances in light and shadow. But under fatigue, the image begins to blur; the lens fogs. Shadows deepen into noise. What once held texture becomes flat and indistinct. Meaning drains from the frame.

Nervous exhaustion activates the R-complex: the primitive survival core of the brain, keeping the body locked in a chronic fight-or-flight state. This triggers a cascade of symptoms, including:

Panic disorders

Cognitive fog, perception thinning, and fatigue

Heightened sensitivity and diminished pain thresholds

Mood swings and detachment

Depression and manic states as the body overcompensates with adrenaline

In seeking relief, people often turn to a wide array of self-help methods, medications, and therapeutic solutions, many of which may offer temporary relief but fail to address the core issue: the mind's locked-up state. Fear of recovery due to relapse can further perpetuate this cycle.

The Problem is: The Mind Locks Up, and blood flow is restricted.

Just like a muscle cramp that tightens to protect against an injury, the body and mind "lock up" in response to exhaustion and trauma. This protective mechanism, while helpful in short bursts, becomes harmful when prolonged. When the nervous system is constantly under strain-whether from stress, trauma, or chronic fatigue-it begins to operate in a defensive mode, leading to mental and physical rigidity.

For instance, imagine a loud noise triggering an immediate physical reaction, people hunching over, tensing their muscles, or freezing instinctively to protect their vital organs. This is a short-term survival response designed to shield us from immediate danger. However, over prolonged periods, the nervous system continues to hold onto these protective states, leading to chronic tension, emotional suppression, and mental fatigue, such as in cases of PTSD and burnout.

Other examples of the body locking up include

Emotional Shock Response

When receiving bad news or experiencing a traumatic event, people often describe feeling "frozen" or "numb." This is the mind's way of shutting down to avoid overwhelming stimuli, but if unresolved, it can lead to long-term emotional suppression.

Chronic Fatigue from Infections and Illness

When the body fights off infections such as viruses, bacterial infections, or autoimmune conditions, the immune system demands a significant amount of energy. This prolonged internal battle can lead to nervous system depletion, leaving individuals feeling physically and mentally drained. Over time, the body may adopt a "locked up" state, conserving energy by restricting movement, reducing cognitive function, and slowing down recovery processes. If left unchecked, this fatigue can become chronic, leading to conditions such as post-viral fatigue, where even minor activities feel overwhelming, and the mind struggles to regain its former clarity and resilience.

Chronic Stress Holding Patterns

Many people unconsciously hold tension in their shoulders, jaw, or stomach due to chronic stress for various reasons. This prolonged state of muscle contraction can lead to headaches, digestive issues, and overall exhaustion.

Fight, Flight, or Freeze Response

While the "fight or flight" response is well-known, the "freeze" response often goes unnoticed and is prolonged. This can manifest as physical stillness, mental dissociation, or emotional detachment as the nervous system attempts to protect itself from further harm.

Nervous Exhaustion in Sleep:

Even in sleep, the mind and body remain in a heightened state of tension. Many people report waking up feeling more exhausted than when they went to bed because their nervous system never fully recovers during rest.

Over time, these locked states of tension can create a vicious cycle, leading to heightened sensitivity, brain fog, diminished resilience, and an inability to cope with even minor stressors. The body's natural repair processes are hindered, and without intervention, individuals may feel trapped in a state of perpetual exhaustion and anxiety.

Unlocking the Mind and Body

To reverse this locked state, we must first understand it, and then consciously and systematically "unlock" the mind and body. By using targeted physical techniques, such as scalp and skull base releases, deep breathing, and visualisation exercises, we can begin to restore blood flow, oxygen, and a sense of safety to the nervous system. Recognising the mind-body connection and addressing these protective patterns with consistent effort is the key to long-term recovery and resilience.

Step 1: Restoring Blood Flow and Oxygen

A depleted nervous system results in poor circulation and oxygen flow to the brain, exacerbating symptoms like cognitive fog and fatigue. This rigid body state restricts blood flow to the mind. Restoring these physiological functions is the first step to healing.

Techniques to Improve Blood Flow

Scalp Loosening Exercise

Take your hand and grasp the top of your hair and crown.

Firmly pull left to right and forward and backwards about 10 times.

Repeat this every 30 minutes for at least a week.

This encourages scalp flexibility, allowing for improved blood circulation and mental clarity.

Occipital Release Technique

While lying down or standing, place your dominant hand behind your head.

Use your fingers to gently pull up at the base of your skull.

Breathe deeply 5-10 times while holding the stretch.

Do this at least every hour to maintain blood flow and oxygenation.

Combined with deep breathing, these techniques help repair the mind by reducing physical tension and improving nutrient delivery to the brain.

Step 2: The Unlock Method – Expanding the Mind

Understanding the R-complex, Ball

Imagine a "**ball**" at the back of your mind, symbolising the *R-complex*. This area is responsible for survival instincts, fight, flight, or freeze. Most people unconsciously clamp down on it, reinforcing their stress and exhaustion.

Your goal:

1. Use your consciousness to expand this ball at the back of your mind, rather than squeezing it.

2. Be mindful of moments when you feel tension at the back of the mind and neck, and mentally release it.

3. Practice this consistently throughout the day to "unlock" your nervous system.

Unlocking = Recovery.

By remaining conscious of this process, you are aiding both your body and mind to relax and repair, breaking a cycle.

Step 3: The Placebo Effect – An Understanding Of The Unlocking Healing Potential

The placebo effect is one of the most fascinating yet widely misunderstood phenomena in medicine. Time and time again, studies have shown that belief alone can drive real physiological changes, easing pain, calming anxiety, even sharpening cognitive function. Acupuncture, ultrasound therapy, and antidepressants have all been scrutinised, with research suggesting that much of their efficacy may come down to expectation rather than the treatment itself. Even techniques like transcranial stimulation and certain forms of talk therapy seem to work, at least in part, because the mind buys into the process. The implications are enormous: healing, to a surprising extent, is something the brain is capable of orchestrating on its own.

How the Unlock Method engages or utilises the placebo effect fully

Instead of dismissing the placebo effect as a mere anomaly or as if it were some kind of illusion, we should recognise it as a powerful force in healing. The "Unlock Method" I have refined deliberately taps into this process. Take the simple example of someone swallowing a sugar pill and feeling real relief. This isn't trickery; it's the subconscious mind partially *unlocking* its own ability to heal, subtly guiding the psyche and body toward restoration.

Step 4: Recovery and Movement

After consistently applying the Unlock Method for a week or two, you should begin experiencing noticeable improvements in mental clarity and energy levels. At this stage:

Incorporate light walking. Walking promotes circulation and oxygenation, which are crucial for brain health.

Gradually increase activity over time, focusing on aerobic exercises.

Allow your appetite to guide you; good nutrition and hydration will naturally become more appealing as recovery progresses.

Avoid overanalysing problems. Prioritise recovery before tackling life's challenges. A healthy mind can naturally resolve many issues.

Recovery is a gradual process; respect it, walk before you can run. Trust in the body's ability to heal with consistent application.

Step 5: Kindness - A Step Toward Greatness

I will incorporate alignment, understanding, and knowledge to aid in recovery. Kindness is often misunderstood; it's not about being "nice" to please others but about maintaining a centred, authentic state that fosters personal growth. I understand kindness has been preyed upon by the weak, so in this instance, I ask you to apply it to yourself to aid in recovery.

A crucial element in recovery is kindness, particularly toward yourself. Kindness calms or tames the R-complex, reducing instinctual reactivity.

Improves emotional regulation and cognitive function.

Strengthens the integration of the mind, fostering balance and critical thinking.

Strengths foundations are not arrogance or aggression; true strength lies in balance, depth and self-understanding, where kindness is fundamental.

Practical Exercise to apply kindness to aid in recovery:

Repeat the thought and feeling: "*Kindness and safe, outside.*"

Feel safety flowing down both sides, or at the sides of your body.

"Outside" reasoning. It subconsciously focuses your mind on nature, remembering that your body is part of it, created from it.

By practising kindness, firstly to yourself, you create the mental conditions necessary for long-term resilience, strength and well-being.

Step 6: Utilising Feeling and Imagination

Your imagination is a powerful ally in recovery. Utilising visualisation and feel techniques can also further assist in healing:

Recall a past moment where you felt completely safe and at peace.

Focus on this memory and associate it with the feeling of kindness and security. Safe, home, outside, as an example.

Repeat it daily to reinforce a sense of stability and well-being.

By engaging both feeling and imagination, you enhance neuroplasticity, making it easier for the brain to form healthier patterns.

Final Thoughts: Trust the Process

The Unlock Method provides a structured, logical and cohesive framework within neuroscience principles. Yet it's a flexible approach to overcoming nervous exhaustion and regaining control over your mental well-being. The key takeaway is that the brain and body, when "unlocked" from a defensive state, can initiate an ongoing natural healing process.

Healing does not have to be complex; trust in the body's natural ability to heal. Also, many issues you may think you have will naturally be resolved in recovery itself.

Recovery, at times, can feel like navigating a corridor, an apt metaphor for the healing journey. As you heal, the past, those old scars, may rise up again, like shadows creeping back into the light. Instead of dwelling on them, try to see this as part of the healing process, a sign that you are returning to a healed state, coming back to who you were meant to be.

Key reminders:

Consistency is crucial. Apply the method daily.

The mind must be used to heal the mind. When it locks, the body follows. But with understanding, with technique, with coherent structure, you can turn the key and follow the road to recovery. Be kind to yourself. Recovery takes time.

This isn't a sprint, it's a return. Respect your pace, not your impatience. Stay present. Focus your attention on the process of healing, not on the fear of relapse. Recovery unfolds in real time, not in regret or anticipation. The method remains accessible under any circumstance.

Avoid self-sabotage. It often emerges from nervous exhaustion and inner frustration. Recognise it not as failure, but as a signal that your awareness is returning, and that you're still on the path.

Psychotropic medications

I will address the ongoing rise of psychotropic medications and their reckless distribution, primarily for corporate dollars, combined with a deterioration of mental health in society. There are medications that, in some circumstances, may, combined with a holistic approach, have a benefit. But this is my opinion based on my knowledge.

I personally have run trials of Antidepressants on myself, to experience their effects and to come to conclusions, as it's a humanitarian issue. Psychotropic medications are like wrapping the mind in a type of plaster you never take off, and some are akin to doing brain surgery with a mallet. I won't go into details, other than there needs to be a thorough examination of the industry.

Take benzodiazepines, for instance; they can provide genuine relief in the short term, but beyond two weeks, they start to create a problem far worse than what you started with. Their potential for dependence and withdrawal complications cannot be overstated.

Rather than exhaustively detailing these issues here, I would encourage anyone interested to watch a concise yet insightful documentary: *How Big Pharma Makes Healthy People Sick* by ENDEVR. I have no affiliation with the creators, yet they have good content, and I believe their documentary presents a broad and *real-world* perspective on the issue. Illicit drugs? They take your soul and follow you to the grave. Even the so-called top-shelf 'fire water' can lead you down the same path if taken without restraint. For those considering tapering off psychotropic medications, or other mind altering sunbstances. The *Unlock Method* may offer a valuable tool to support their journey towards recovery.

If people wish to post on social media that the "Unlock method" has helped, please do -TR.

Chapter 20
Alignment

Where do I begin? When I was young, I learned that self-understanding wasn't just about looking inward; it fostered a deeper appreciation of the world beyond myself, the real world, which was ultimately far more interesting than the particulars of my own story. I never cared much to talk about myself because everything out there held a greater fascination. I've always carried this balance; it could be worse, but it could be a lot better, too. My childhood? It wasn't easy, but I can't complain.

I grew up in that in-between space, close enough to the city to see it, but far enough away to belong to the wilderness. After school, I'd disappear into the land, running with my best friend, nature, and my other best friend, a dog who was more freedom than fur. We roamed an ancient stream, tracing the stories of fossilised rocks, each one a reminder that this mountain I called home had once risen from the sea. The landscape had its own way of teaching, about a timeline far older than human existence. Earth's time, not mankind's.

Looking back, I realise my mentors weren't always the traditional kind. There were the older boys, young men, really,calm yet strong, who embodied a quiet strength, a mix of adventure and rational restraint. And then there was another influence, a young woman much older than me, whose kindness and wit were matched by a deep sense of compassion.

Money was tight, so we had a lodger in her late thirties. She saw things others didn't. She filled the house with music, the kind that lingers in the background but teaches more than books ever could. Those songwriters, I realise now, were prophets in their own right, tapping into something deeper, translating it into chords and verses that made sense of things words alone couldn't. Many of those songwriters, I now recognise, were deeply attuned to the human condition, modern-day prophets, able to channel something profound through their art.

I was lucky. Surrounded by such influences, I matured quickly. But that kind of growth isolates you. School felt distant, as though I was moving at a different rhythm. Immaturity was the currency there, and I never had much of it to spend.

My greatest refuge was my grandmother. Two hundred or so miles away, but always close in the ways that mattered most. She called us grandchildren, her little people, and in her world, laughter and kindness were as constant as breathing. But don't mistake kindness for softness; she was sharp, nobody's fool, and she carried a strength that didn't need to shout to be heard. Summers with her were a sanctuary, a reminder that real strength doesn't need to posture. It just is.

My sixth-grade teacher wasn't just a man in a classroom. He was something more, my first real model of intellectual honesty. A force. A man of principle, the kind that didn't need to be spoken, because you felt it. Every student, treated the same, no favourites, no lesser ones. And because of that, we all respected him. But what set him apart wasn't just his moral fibre; it was his passion for cosmology, not something he taught, but something he was. It wasn't a subject; it was a current moving through the room, carrying us with it. Not fantasy, not stories spun for comfort, this was real-world magic, the vast, intricate, breathing universe. And because of him, I started searching for my place in it. The truth spoke for itself, and my scientific curiosity was ignited.

Science called to me, but so did the birds outside the classroom window. They were free, wings wide on an endless holiday, creatures moving through the sky while we sat confined, quite literally, within walls, dissecting frogs in an attempt to understand life. The irony was not lost on me: we were being taught to analyse living things by taking them apart, while the deepest insights often lay in simply observing life as it was. But no labels could truly explain it. And something in me knew, this education system, as it was structured, did not necessarily cultivate the kind of intellectual freedom I sought. If anything, it often stood in the way.

My first real encounter with love, and with truth in its rawest form, did not come through books or whispered myths. It came in the form of a girl, seventeen, living across from my grandmother's house. She was sharp, quick-witted, and bold in ways that made conversation an art form. And yet, behind all that, there was a softness, an intelligence that moved beyond mere intellect into something more elusive. Where she was mercurial, I was the ground, catching the charge, steady but never still. We spoke in rhythms, in laughter, in glances, something beyond thought, deeper than time. The mind gave way to something vaster, to the tender ache of love's growing pains, of a deeper knowing. Heart to heart. An earth-moving love that brings you to your knees, where words fail but the heart understands, where silence speaks louder than sound.

Love, real love, was no gentle, flowing stream. It was a river, fast and wild, throwing you into rapids, making you hold on tight, never knowing if you'd sink or make it through. A current that pulled between pain and bliss, that left you breathless, drowning and gasping, only to hurl you onto golden shores where the heart cracked open, not from breaking, but from something too vast to name. And if I wrote a book about those days, it still wouldn't be enough. Love like that isn't meant for pages. It lives in the spaces between words.

But in the way of things, every story begins, every story ends. That's the cycle. And the distance stretched, her place to my mountain home, then further still. She moved, her family moved. Love didn't die; it retreated. Like the tide pulling back, taking everything with it, leaving me standing in the shallows, empty. But the pain I felt wasn't caused by love itself. Truth, through reason, whispered to me: it's not love that wounds, it's the absence of understanding love that cuts the deepest.

As someone inclined towards observation, I took this as an opportunity for reflection. I studied the nature of love in the way an astronomer studies the night sky, patiently, methodically, without assumption. And what became clear was that love and truth were, in some way, inseparable. They moved together, woven from the same thread. Love wasn't something we owned; it was something that moved between us. A current, a force, something alive. And in that flow, there was clarity. Love induced intelligence, sharpened the senses, and made the self-deeper, stronger. The more I saw, the more I understood: truth and love weren't separate things. They were two sides of the same motion, the same energy. A duality. It was like stepping into a version of myself I hadn't met before-but that had always known me.

So I followed, truth. Even when love wasn't clear, the truth was the path. It had to be. Day by day, year by year, I let it guide me. I asked myself the hard questions, stripped away illusions, and stood firm in what was real. I committed myself to honesty, both with myself and with the world. My principles took root, but not like stone, like a river, moving, strong but fluid, bending with understanding. This was not a process of rigidity or self-righteousness, but one of fluidity, guided by compassion. And if there's one thing I learned, it's this: never grow old, only grow wiser. Wisdom is the process of moving beyond the reflexes of immaturity and into a deeper, more integrated self. But the child at the core? That stays. Because innocence is not weakness; it's genuine clarity. And kindness? Kindness is a step towards greatness, a necessary precondition for higher forms of intelligence. A mind without kindness is a storm with no peace. Kindness is taming the instinct, harnessing the will to achieve balance; that's where true power lies. The ability to stand firm in yourself, to walk your own road.

And through that, I found love again. Not in another person, but in its vastness, an ocean I could step into whenever I chose. And as I aligned myself, something profound happened. The mind deepened, both aspects growing and coming together, a marriage of the mind and truth, moving as one, binding me in a deep sense of wholeness.

If we view the mind as a system of alignments, then truth serves as its axis. And the psyche, when properly aligned, becomes something greater than the sum of its parts. One could think of it as a kind of portal, a gateway to deeper dimensions of experience. This is not a metaphor in the trivial sense, but a recognition of how consciousness itself is capable of functioning. The mind, when fully engaged with truth, can perceive reality in ways that are not constrained by conventional limitations. And in this, we find the true power of alignment, not as an abstract idea, but as an experiential reality.

We can think of the unconscious mind as we would an acorn: within it lies the blueprint for a great oak, just as within each of us exists the potential for greater understanding. In the same way, a person carries their potential, even if they sleepwalk through their days, lulled by the hum of artificial distractions. But truth, real truth, grounded truth,is the soil that awakens the seed. If you choose to root yourself in the substance of things, in the undeniable, you begin to grow. And growth isn't mystical; it's natural. Grounded in evidence and reason, we develop into something greater.

Intuition, when properly understood, is not mystical. It is a form of deep pattern recognition, what we might call "deeper common sense." What once seemed an inexplicable gut feeling in childhood often becomes obvious in adulthood. That's intuition revealing itself. It's not guesswork, it's understanding before explanation. And when you build on that with reason and critical thought, you sharpen your mind. You start seeing through the noise. Coherence is the key; apply common sense. The more sense you make of things, the stronger your intelligence becomes. And intelligence isn't something handed down by fate or some cosmic lottery; it's creation's gift. The idea that certain people are "chosen" is, in my view, deeply misleading. Those who choose to use what creation has given them are the ones who become chosen.

I don't dress "Alignment" in mystical robes. I don't wrap it in vague spiritualism. I speak in analogies because knowledge needs a way to be carried. I could talk endlessly about the nature of a chair, but to truly grasp its purpose, you must sit in it. Alignment is the same. It's not about rituals or titles, it's about trust. Trust in yourself. Don't underestimate your own existence. Your origins stretch back beyond time's memory; your intelligence is the inheritance of billions of years of life's unfolding story. The atoms in your body were forged in the hearts of dying stars. Formed by the tides of evolution, you stand here now. Don't squander this extraordinary inheritance on the superficial waste heap.

Our grasp of consciousness is still in its infancy. Consider the documentary Brain Man: The Boy Genius with the Incredible Brain, it provides a real-world glimpse into the extraordinary capacities of the mind: intuition, intelligence, and connection. We're standing at the edge of something we don't yet understand. And beyond that edge, alignment stretches into dimensions we haven't mapped yet. The more intelligent life becomes, the greater its capacity to interact with higher-dimensional realities, whether consciously or unconsciously. I won't go too far into details here, that's for the Theory of Everything to unravel. The work must be solid, tested, and proven.

Heaven is up, but we fall into love, not up, but where is "in"? Love isn't just a feeling, not just some emotional state. When viewed through a scientific and dimensional lens. A fluid-type substance, a force intertwined with the truth. And perhaps what we call grace, from a scientific perspective, is akin to gravity. A quantum reaction in the mind. If life exists in three dimensions, why does it seem to extend beyond them? Maybe we are marginally in the fourth, and as intelligence expands, so does our capacity to interact with it. Consciously or unconsciously, it moves in us. But it must be studied. Hypotheses are warranted, but no leaps of blind faith, only rigorous, grounded exploration will prevail; pseudoscientific conclusions must be avoided entirely.

And here's the hard truth: where humanity is and where it should be, there's a chasm. A vast canyon carved by neglect and distraction. And yet, so many see it. Those who think clearly, who strip away the nonsense, who trust their common sense, they recognise history looping back, repeating in the worst ways. A civilisation that, despite living in an era of unprecedented access to an ocean of knowledge, chooses to swim in pools of ignorance. That's a dangerous game. But the truth, if enough people choose it, can intervene. And science, a self-correcting process, must be applied. Because if we don't wake up, if we don't shift course, humanity's evolution won't just stall, it will spiral. And we'll find ourselves sliding not forward, but straight into catastrophe. The Twilight Zone of our own making.

I walked the solitary road for years, not for escape, but for truth. The kind that doesn't bend to convenience. Clearing the path for human intelligence, that's the work. Not for glory, not for gain, but because the gift is in the giving. If we don't recognise our worth, we remain lost. But when we do, when we truly see, we advance.

Does alignment mean becoming some kind of monk, some sage on a mountain? No. Alignment doesn't strip you of your individuality; it sharpens it. It's about living with depth, whatever that means to you. Maybe it's raising a family, perhaps it's setting sail alone. Maybe it's both. It's yours to define.

And true friendships, those are the family you choose. The deeper the mind, the deeper the understanding between people. A kind of unspoken language, a recognition. No need for conflict, only dialogue. No need for division, only clarity. Because truth isn't at war with truth, it resonates, and through it, we rise.

Under this banner, we move as individuals, yet as one, bound by reason, forging a shift in consciousness for the good of all and a deeper respect for universal life.

The wisest step forward in these times is to step back. Many intelligent people find themselves physiologically depleted, physically drained, and mentally worn. They need space to heal, to reclaim themselves. The priority must be recovery, withdrawing where possible from the noise of social media and the relentless grind of a world slipping into dystopia. They must be kind to themselves, find solace in the simple things, keep their heads down, and stay safe.

But for those with strength, for those who stand firm in themselves. I ask for your focus, your presence. I need to anchor myself to the earth, to the fundamental nature of things, as fully as possible. Only by doing so can I fully bring my abilities to bear in a way that truly serves you.

Trust the process. Progress happens one step at a time.

Carl Sagan. Pink Floyd, On the Turning Away.

Chapter 21

Alignment over Obedience

Religion

I was never one to subscribe to religious or spiritual belief systems. At best, I was interested in whether they held any practical insights, something real hidden beneath the layers of rules and jargon. And who can claim ownership of God? Such things defy reason.

One of my earliest memories of religion was around the age of five, sitting in a church and staring at a man, bloodied, broken, nailed upon a cross like some discarded soul from the street. I couldn't understand why the adults simply sat there, watching. Saying their prayers while he hung there suffering. It made no sense. So I did what made sense to me. Walked up the aisle, tried to help that man down, convinced that if I could just get him to my grandmother's house, she would clean and dress his wounds, make him better. The preacher was not impressed, but my grandmother? She was proud. Because, like most children, I had compassion and the instinct to act, before we are taught to obey, to look away, to become dulled out and indifferent.

And I couldn't buy into the story they were selling. The idea of an all-powerful being playing some cosmic game of chess, creating 'monsters' just to test us, seemed deeply flawed. And the notion that my dog, one of the purest, most loving creatures I knew, wouldn't be welcome in heaven? That sealed the deal. But Jesus himself? A man who flipped tables, who stood against those who hoarded and controlled, who called out the thieves in high places? That part, at least, made sense, along with words of love and compassion.

I work from evidence. So, whether or not Jesus walked this earth is another matter. But the story, what it means, what it carries, perhaps that's the real question. Did his love open hearts blinded by fear? Did he bring healing, not to the flesh, but to the sickness deep within, the kind that turns hearts to stone? Did he step beyond this reality, see through the thin veil, touch something fluid, something alive beyond the limits of ordinary perception? I don't know. Without a time machine, the truth remains out of reach; all we have are stories passed down, lingering in the realm of speculation.

But if people find something in it, if the myth of the man holds them together, if it keeps them warm in the cold of this world, who am I to deny them that comfort? If it grants them strength rather than shackles, if it moves them towards kindness rather than cruelty, then let them have it. And for those who have been psychologically shipwrecked, who experience life as a kind of perpetual drowning, it is entirely understandable that they would grasp for anything that offers relief. I won't ridicule them. I won't shame survival, as long as it does not bring harm to others.

Consider Martin Luther King Jr., one of the most profound moral voices of the last century. A man who stood for truth, who carried a heart too strong for the system to stomach. He saw through the game, and he called it out. He walked it, lived it, and breathed it. And for that, they silenced him, with a bullet, with fear, because he dared to remind people of their own power. He was dangerous not because of his theology, but because he articulated a vision that threatened the foundations of power. As he put it in 1967:

"We must rapidly begin the shift from a 'thing-oriented' society to a 'person-oriented' society. When machines and computers, profit motives and property rights are considered more important than people..."

He threatened their order, not with his religion, but with his humanity.

In his later work, he placed increasing emphasis on economic justice as a foundation for true equality. In his 1968 essay, he called for an Economic Bill of Rights, a vision that included guaranteed income, affordable housing, and access to quality education. King understood that without economic dignity, civil rights for all remain unfinished business.

And if that's what his faith gave him, if that's how he used it, how could I, in good conscience, mock it? When he wielded it so rationally, so effectively? When it served as a force for justice rather than for power, delusion, or wealth.

I see the truth in what clear-minded agnostics and atheists say. The hypocrisy, the destruction that religion has left in its wake, it's there, written in blood and history. That said, there's little value in dedicating too much time to deconstructing it here. From my perspective, truth, grounded in facts and evidence, is a religion in itself, the purest form of reality. It doesn't need temples or holy books. It doesn't ask for blind belief. And in light of that, traditional religious doctrines seem... Small. Arbitrary. Inconclusive.

Is there a God? From my personal perspective and experience, yes. But not the kind that watches, judges, or whispers commandments into the ears of men. Not the kind sold in churches or weaponised in politics. It isn't easy to articulate, yet undeniable in its presence. God has never spoken to me. But I feel it. It's not something to bow to, not something to worship. It's alignment. God just feels like normal, like more normal than normal. A deep, undeniable, strong sense of home with profound appreciation. It just is.

But no, I am not one or fully connected. That kind of weight would crush me. It is more a matter of aligning with something beyond myself, of receiving a reflection of it. I can extend that reflection outward, but not everyone is ready for that. Some people, if they are not balanced, if they are not strong, may find it overwhelming.

At some point, I will guide others in achieving this alignment. It is not difficult, it will not require faith, and it will not come at any cost. My reasoning is simple: the more individuals who stand grounded in themselves and align, the better.

But I don't ask for belief. I expect scepticism. I would expect anyone to be highly sceptical of such claims unless they had personal or verifiable evidence. As for dimensions, other layers of this reality, these remain ongoing areas of study, challenging enough in themselves without unnecessary mysticism.

Defining God is no simple task. In fact, it is a near-impossible one, given that our understanding of life itself is, at best, rudimentary. We are beings composed of the fundamental building blocks of the universe, particles and forces that we barely comprehend. To quote Mahatma Gandhi, "*There are people in the world so hungry that God cannot appear to them except in the form of bread.*" This speaks to a deeper truth, a hunger not just for sustenance, but for clarity. It highlights the impossibility of distilling God into a singular, coherent definition.

The question is not whether believing in God is right or wrong; that is the wrong question. Rather, the issue lies in the attempt to define God with certainty. For fallible human beings to claim, "*This is the word of God,*" with absolute conviction, is to take an extraordinary leap, one that risks turning truth into a proprietary claim. To assume that the vast and unfathomable can be neatly contained within the pages of a single text is to risk writing one's own doom the moment ink meets paper.

The notion of divine communication presents an equally complex problem. If an intelligence far beyond our comprehension exists, how would it communicate? Could love be contained within a single syllable? Love is something you feel, something that moves through you, something that becomes. You cannot hold it. You cannot break it down into letters and expect to decode it.

And why doesn't God intervene? This question has haunted human consciousness for millennia. But consider this: does David Attenborough interfere with the natural world he observes? Does he poke animals with sticks? Does he control the rivers and the trees? No, he observes. He lets nature be nature. Likewise, those who cherish national parks and wildlife sanctuaries recognise the value of letting nature take its course. Perhaps what we call God does the same, allows existence to unfold, and refrains from interfering with the fragile balance of what is. And perhaps that is the problem. Perhaps we keep waiting for something outside of ourselves to step in, when all along, the power has been in our hands.

The temptation to invoke mysticism in the face of the unknown is strong. But quantum mysticism, the vague, unscientific co-opting of quantum mechanics to justify spiritual claims, is a distraction. Progress is not made through wishful thinking or poetic metaphors; it is achieved through rigorous experimentation, empirical data, and intellectual honesty. Science, not superstition, is the only viable path forward. The first scripture was not written on paper; it was written in the stars, in the land, in the movement of the cosmos. In this sense, science has functioned as humanity's safest attempt at religious inquiry, one that began not with dogma, but with curiosity.

Science is a self-correcting process. It is not perfect; it struggles, it stumbles, and at times, it is deeply challenging. But it remains the only framework we have for approaching truth with humility and precision. And if defining God is possible at all, it will be through this process, not through doctrine, but through discovery.

And finally, if God exists, it is irrational to assume that such an entity would favour one person or people over another. Would an omniscient being ignore innocent children buried beneath blood-stained rubble while bestowing favour upon those who claim divine authority? Ask yourself, who gets to claim God's protection? Who gets to wield it as a weapon? History answers. "Who says I am not under the special protection of God?", Adolf Hitler. So tell me, who is defining God? And who is merely using the word to hide what they truly are?

Agnostic, Atheism

Einstein, Carl Sagan, Roger Penrose, and many others, rational agnostics, have expressed openness to the possibility of a higher power, not out of superstition, but because of the sheer depth, magnitude, and complexity of the universe. The cosmos is vast, rich, and awe-inspiring, and within it, life emerges in all its beauty, like the crown of the universe itself. It is this very grandeur that compels us to wonder, to question, and to strive for understanding.

The Book of Life, older than gods sculpted by men, is a narrative that stretches back billions of years, ours to decode. Regardless of whether we can ever fully comprehend the true origins of existence, our responsibility is clear: we must seek to read this book with intellectual honesty and curiosity, to know the difference between what is real and what is merely desired. Science provides us with the tools to navigate this inquiry. Until we have evidence, we must treat the idea of what's beyond our comprehension as a hypothesis, not a given fact.

And to the devout atheists, I see you. You hold your ground, rejecting belief in anything beyond what can be observed, tested, and proven. Your commitment to rationality, empirical evidence, and common sense is admirable. I respect that. But let me ask you this: Do you see a religion that does not call itself religion? Do you see the temples built from profit, the altars of industry, the gods of wealth demanding your worship? Are you certain you are not serving a material religion of your own? Is your nirvana the acquisition of wealth, where more than enough is never enough, or the glorification of status? Is this not a form of idolatry, an obsession with accumulation, just another brand of worship? Haven't both atheism and theism, in their more extreme forms, been responsible for untold destruction throughout history?

Do you not see that irrationality is not confined to religious dogma alone? That the modern obsessions, materialism, consumerism, and unchecked ambition, can be just as corrosive? The rejection of faith does not grant one immunity from delusion. One can abandon belief in deities and still find oneself devoutly worshipping capital, reputation, or ideology. There is no moral high ground in renouncing one form of blindness only to kneel before another.

More often than not, the sanctimonious are perched not on reason but on the mount of hypocrisy. Pointing out the flaws of organised religion requires very little effort, perhaps even less insight. That entire industries of podcasts, book tours, and lucrative lecture circuits have emerged around this effort should raise a reasonable eyebrow. Is it about truth? Or is it another theatre of profit and self-confirmation?

Unless, of course, I'm mistaken, and they quietly funnel most of their gains to those living in the shadows, the vulnerable, the discarded, the disillusioned. But I doubt that's the case.

In truth, those who position themselves as moral arbiters, claiming to expose religion's corruption, often end up profiting from the same illusion; they've simply inverted the altar. They haven't transcended the dogma; they've monetised it.

The truth, if it's anything at all, requires a different posture. It doesn't stand atop podiums or bask in applause. To grasp it fully, you must kneel. Not in surrender to dogma, but in recognition of something older than argument: humility. A humility that isn't weakness, but a strength forged in silence.

A balanced mind does not serve gods, wealth, or ideologies. It seeks clarity, endures contradiction, and, above all, learns to stand, even in the unknown, with honesty.

If we had the courage to fully embrace reason, even just the clarity of common sense, the grip of our greatest crises might begin to loosen. Not over centuries, but perhaps by tomorrow. But let me be clear: this is not about defending one belief system over another. I have no interest in smuggling God in through the back door or reigniting tired theological skirmishes. I speak only from experience, what I've lived, not what I demand others adopt. No dogma. No conversions.

We don't need another round of who's right and who's deluded. That narrative is exhausted. It's been playing on loop for centuries and hasn't saved us.

What we need is something quieter but infinitely more potent: self-examination, critical thought, and a willingness to look inward without flinching. These are the tools, not inherited doctrines, celebrity philosophers, or bumper-sticker scepticism.

These are what carry us to a clearer, more honest understanding of who we are, and where, if anywhere, we might fit in this strange, unfinished universe.

Spiritualism

Spirituality. It has been diluted to the point of meaninglessness, dressed up in soft words, made weak, bought and sold. I don't use the word much. How far do we take it? Are we to tiptoe barefoot across the grass, paralysed by the possibility of cutting short an ant's existence?

Suffering is a part of existence; there is no escaping the reality of it. To deny this is to remain intellectually stunted, unable to harness the full capacity of the mind. If finding a rational balance is too difficult, the individual is left at a loss, adrift in a sea of bad ideas, easy prey for those who build illusions and sell them as reality.

And then there are the ones who say life isn't real. Reality is an illusion. But if that's true, then what is-, feeling it? What is hurting? If reality is an illusion, then what, exactly, is experiencing it? The only illusion to reality is the surface, to the depths of reality.

Religion. The uniforms of devotion. The cloning of identity. I don't know if it's fear or conditioning, but when everyone looks the same, talks the same, thinks the same, it stops being about individual growth and starts being a system. And systems aren't built for freedom. They're built for control, even if it's passive.

As for meditation, I have never found it necessary to assume a lotus position or chant in order to cultivate deeper clarity. If you think too much, you lose touch. It's the mind running low on biological energy. The mind, like the body, can exhaust itself. I trained myself not to think unnecessarily, and then taught myself to think only when it served a purpose, whether for reflection, observation, or deliberate action. What I found is that a well-rested mind isn't just more functional, it's far more capable of realising its potential. You can remain grounded, act with clarity, and still give the mind the rest it quietly needs. The psyche is an instrument. It benefits from discipline and from cultivating the rare skill of simply not thinking. Like anything else, it takes practice.

As life is a wave, you either learn to move with it or you drown under it. Balance isn't stillness. It's knowing when to shift, when to stand firm, and when to let go. When the mind is optimised, when thoughts are disciplined rather than left to run unchecked, one can enter a state of flow. Unlocking the depths of your mind is not a mystical endeavour. It is a matter of attentiveness, of discipline combined with thought management. The back of the mind is a door that's never locked. But you have to want to open it fully. And you have to be willing to take care of what's inside.

And then there is the idea of absolute "enlightenment", the pot of gold at the end of a rainbow that was never there to begin with. But the ones selling it? They're the ones who get rich. It is a game of mysticism, a racket of grand promises and carefully crafted illusions. But there is a path of appreciation with alignment, one that never ends, where you get to be a child again, wrapped in wisdom and strength. It is a path so fulfilling that one might choose to return to walk it again without hesitation. No big deal.

But what of those who deal in esoteric swami nonsense, speaking in riddles? Word magicians, who give the illusion of profundity? It is no different from the skill of an artist who paints an incomplete picture, allowing the viewer's mind to fill in the missing details. The result is a sense of personal revelation, an illusion of deep understanding that is, in reality, sleight of hand. But for the artist, it's a sharing hand. This technique has been exploited for centuries to sell spiritual woo "wisdom," often accompanied by lucrative retreats, expensive courses, and adoring, starry-eyed followers.

Perhaps I am being too critical. Because I see an industry built on vague promises, convincing well-meaning people to become perpetual seekers, never realising they are chasing their own tails. There is a constant supply of people, disillusioned or exhausted, those desperate to escape a manufactured reality that is, indeed, difficult to cope with. And while it is each individual's responsibility to arrive at their conclusions using reason, one can only hope that, for those who feel lost in this world, love, or something resembling truth, will find them, free of charge.

Progress?

Consider whales. They do not pay mortgages or rent, yet we, the self-proclaimed pinnacle of intelligence, believe that perpetual economic servitude is a sign of advancement. If intelligence means constructing a system where shelter, one of the most basic needs of any living being, is treated as a privilege or dangled as a dream, rather than upheld as a human right, then maybe it's time we rethink what it means to be intelligent.

How did we let this happen? It did not come all at once. The slow erosion of the right to a home is a case study of how dysfunction can become normalised. It did not vanish in a single act of policy or conquest but was gradually and systematically redefined, one law here, one lie there. A shifting of values, a bending of truth eroded by centuries of economic conditioning, manipulated desires, and fears of scarcity. Like the hour hand of a clock, imperceptibly shifting until one day, the rational becomes radical, and a roof over one's head is not a given but a distant aspiration, a space to exist, turned into a dream.

At some point, we should ask whether humanity has been sleep-walking into an elaborate trap, one step at a time, over generations. A trap so complete we don't even question it anymore, while enduring it is seen as responsible. But this dream is not just American. It evolved into a type of sickness, spreading across the Earth. A sickness where power feeds on the many, where greed replaces need, where the people are kept in a cage so wide they don't see the bars.

George Carlin's observation was right: "*That's why they call it the American Dream, because you have to be asleep to believe it.*"

Comedians, the truth-tellers of our time, with clear observational prowess, speak so we don't forget, to offer momentary relief from the contradictions of our existence. Yet the most unsettling truths are often those delivered with skill in jest. Today, the world appears more like a dystopian satire than a place of rational progress. We are plugged in, tuned out, and, in the most conservative estimate, 95% of humanity functions as little more than a resource, life, for profit batteries feeding a system that has mastered the art of self-preservation at the expense of the individual.

In the theatre of fate, irony has the final curtain call. The Matrix is beginning to look less like science fiction and more like a documentary.

But I do not believe in hopelessness. I do not believe we are meant to stay asleep. Intelligence, when applied correctly, has the capacity to reshape even the most entrenched systems of dysfunction. It is not a matter of whether we can overcome this cycle, but whether we will choose to. The phrase "for future generations" is often wielded as a rhetorical tool, a slogan, like a chant without meaning, used to justify decisions that actively degrade the world we claim to protect.

Perhaps the real leap in human evolution will not be technological but ethical, when we begin prioritising future generations as more than just an applause line.

Rejuvenating and elevating humanity's collective intelligence so that all may live full lives will take generations. But every path begins with a single step.

Morality, Law, and Judgement

Morality is not the product of belief systems; it is woven into the very fabric of our evolutionary path. We see it in nature, particularly in mammals, manifesting as nurturing, empathy, cooperation, standing together, and social cohesion. These behaviours are not arbitrary; they are survival and intelligence, integral to the development of higher reasoning.

Natural laws, too, are part of the evolution of the cosmos and, therefore, life. The same applies to human societies; laws arise as an extension of intelligence, a necessary framework for order, but in balance. The issue is rarely with the existence of laws themselves, but with how they are applied. When enforcement lacks wisdom, it breeds discontent. A more intelligent society, one in which individuals possess greater self-respect, would require minimal law enforcement and fewer external constraints. A coherent mind that thinks freely does not need chains.

Judgment is not just inevitable; it is essential. Without it, we could not make even the simplest decisions, driving a car, crossing the street, choosing whom to trust, or carrying out any number of essential tasks. To think critically, to weigh consequences, is to judge.

But those entrusted with legal judgment must be held to the highest standard. A corrupt legal system is more dangerous to society than the worst injustices committed outside it, because when the law itself loses integrity, so does everything built upon it.

War

What can be said? Atrocities, brutality, and madness still march on, sometimes wrapped in flags, sometimes dressed as necessity. The scale of human suffering defies all reason, and yet we continue to rationalise it.

But this book isn't a catalogue of despair. My focus, our focus, must remain on what can be done. Intelligence must rise where violence has ruled. Awareness must tip the scales toward resolution. If there's any hope left, it lies not in recounting every horror, but in realigning the course ahead.

Politics. Are We Exhausted Yet?

Governments across the world increasingly function as subsidiaries of corporate interests. Their primary role, along with that of mainly corporate media, is best summed up by Garry Kasparov: "*The point of modern propaganda isn't only to misinform or push an agenda. It is to exhaust your critical thinking, to annihilate truth.*"

At the core of this dysfunction lies an undeniable reality: forces of self-interest own the board, move the pieces, and rewrite the rules. Corporations and the ultra-wealthy have amassed an unprecedented degree of control over political systems. Through lobbying, campaign financing, and regulatory capture, they ensure that legislation serves their interests while the costs of society, stagnant wages, rising living expenses, and economic instability, are borne by the average person. This is not a flaw in the system; it is the system. The bureaucracy, the legal loopholes, the influence networks, these are not signs of mismanagement but of a well-oiled machine designed to consolidate power at the top, down to the red tape of corruption at the bottom.

History echoes back at us. The far-right, in all its forms, plays the same tired song: power, control, hierarchy, exclusion. It has frequently been the catalyst for the collapse of empires, often driven by greed, authoritarianism, and a pathological resistance to change. Rome's ruling class, the late Ottoman Empire, fractured under its own weight, and 20th-century fascism followed the same pattern: rigid hierarchies, corruption, economic stagnation, and eventual self-destruction. The real danger is not ignorance, but ignorance emboldened by arrogance. And those who weaponise it, whether politicians, media figures, or corporate elites, know precisely how to do so. The exploited rarely see their exploitation until it is too late, and by then, they are either too invested or too broken to resist.

But the far-left is not exempt from scrutiny. When it goes too far, replacing actual discourse with emotional coercion, it creates a '*cotton-wool*' society, a cage wrapped in good intentions at the expense of intellectual honesty. This fosters an environment where dissent is not debated but silenced, and where inclusivity is enforced through passive-aggressive coercion rather than genuine understanding. The result is a form of soft authoritarianism, ideological enforcement prioritised over truth and human unity, which ultimately fragments society rather than unites it.

This can lead to young people feeling disenfranchised, particularly young men who feel passively coerced, leaving them vulnerable to propaganda. The far right and their acolytes of intellectual mediocrity consciously aim to appeal to angst-ridden youth, serving no purpose other than to empower authoritarian regimes. Yet these acolytes are not merely ideological opportunists; they are profiteers, marketing themselves as the only ones willing to say what others will not, as defenders of 'truth' in a world supposedly gone mad. They are not leading some grand intellectual rebellion; they are monetising discontent, selling grievance as a product while ensuring their followers remain dependent on their rhetoric rather than actually growing beyond it. Their real goal is to disengage young people from genuine empowerment, to redirect their frustrations into a machine that serves their own wealth and authoritarian ends rather than the individual's growth.

If we want to foster a generation capable of resisting manipulation, we must reject both the rigid dogma of the far right and the overprotective stagnation of the far left. The answer lies not in extremes, but in the courage to seek truth beyond them.

The real goal should be to protect human rights in a way that preserves individual freedom and open discourse, not to replace one form of oppression with another. The world does not need more ideological enforcers; it needs people who stand for each other, not over each other.

The Illusion of Democracy

Despite our faith in democracy, the concentration of wealth and power tends to corrupt any system, whether capitalist, socialist, or authoritarian. The only reason democracy remains the best framework is that, when functioning properly, it provides mechanisms for accountability. But when corruption is mostly hidden yet rampant, institutions erode, and the public disengages, democracy becomes little more than a managed illusion.

The ultra-wealthy do not need to destroy democracy; they simply need to render it ineffective. That has already happened in many respects. If people desire only wealth, that is their problem, but there is something deeply wrong with a system designed for the people that allows the wealthiest to extract resources from the working class while contributing little in return. If this dynamic is not corrected, history suggests an inevitable outcome: collapse. Those who sustain the system, workers who move mountains, builders, farmers, are also those with the greatest power to bring it to a halt.

But meaningful reform requires an informed public, and on that front, we are failing. Noam Chomsky has spent decades dissecting the mechanics of power, yet his work remains largely unread by those who need it most. In 2025, he was ranked the world's top public intellectual, a recognition that speaks for itself. He was also voted the leading public intellectual from a list of 100 prominent thinkers in a British magazine poll. If knowledge is power, then understanding is the first step: know thy systems. *Requiem for the American Dream* with Noam Chomsky DOCUMENTARY - Politics, Philosophy. is essential viewing for anyone seeking clarity on politics and philosophy.

The American Dilemma

In the United States, the dysfunction is particularly acute, not because it is unique, but because it was predictable. The last two presidents each won with roughly 30% of eligible voters. The largest, most significant political faction in the U.S. is not Republican or Democrat, but non-voters, people who are either disillusioned, disengaged or convinced their vote is meaningless. If they ever mobilised as a unified bloc, they could redefine the entire political landscape.

The likes of Bob Dylan and Leonard Cohen have spent decades warning of this decline through song, among many others, perhaps because music can articulate what comes from an ancient light, or the depths of the unconscious. Cohen, his words, sharp as a blade, soft as a whisper," Leonard Cohen - *Democracy* (Spoken Word)," not as a celebration, but as an elegy. Dylan, the mapmaker of the times, called it before most could see it—"Bob Dylan - *The Times They Are A-Changin,*". Except change, when not guided, can be just another turn of the same wheel.

Moral Arithmetic: Balancing the Books of a Broken Nation

The Department of Defence. A machine with an appetite too big to account for has consistently failed to meet even the most basic standards of financial accountability, after seven consecutive audits. Seven times, it has failed. Its staggering $824 billion budget moves through its hands like water through cracked earth, and still, they ask for more. If this were any other institution, public or private, such incompetence would trigger immediate reforms. The boards would be wiped clean. But here, the silence is thick. A rational approach would involve independent and thorough investigations by highly qualified auditors, working in conjunction with military experts, to identify and eliminate waste while maintaining its formidable defence force. Transparency and efficiency are not mutually exclusive; they are prerequisites for any institution that claims to serve the public interest.

Meanwhile, the tax system bends itself around the wealthy like a shield, a playground for the ultra-wealthy and corporations, who exploit legal loopholes to minimise or outright evade taxation. In some cases, they pay nothing at all. One particularly egregious strategy, the "*Buy, Borrow, Die*" approach, allows individuals to acquire assets, borrow against their appreciated value to access vast sums of untaxed wealth, and ultimately pass them on to heirs tax-free, their wealth reborn clean. Similarly, the so-called "*billionaire borrowing loophole*" enables the richest Americans to live off borrowed money secured by investments, sidestepping income taxes entirely. These are not examples of ingenuity; they are systemic failures.

While conservative estimates suggest that closing these loopholes could generate over $100 billion in a decade, that's the official number; the real figure is almost certainly far higher. The IRS has reported a "tax gap" of $688 billion per year, and that's just what can be quantified. When factoring in offshore accounts, shell corporations, and hidden streams of wealth, the real number is lost in the shadows, but it is vast. The actual amount of lost revenue likely reaches into the trillions over time. The issue isn't a lack of resources, it's a lack of enforcement. Closing such loopholes would not require new taxes on the ultra-wealthy; it would merely ensure they pay what is already required by law.

At the same time, the United States, the wealthiest country in the world, ranks dead last in healthcare among its democratic peers. This is not an abstraction. It translates into preventable deaths, financial ruin for families, and a country that spends more on healthcare per capita than any other nation while delivering worse outcomes.

But imagine, just for a moment, if the numbers were made to add up. If the Pentagon answered for its spending, if the wealthiest Americans were required to pay even a fraction of what they actually owe, the result would be twofold: the deficit shrinks, resources shift, and suddenly, the impossible becomes possible, the ability to fund essential public services, including a healthcare system that does not leave its citizens behind. A healthcare system that, in many cases, does not crush its own people. These are not radical propositions. This is balance. It is simply what a functional society should, and morally must, prioritise.

Consider this an olive branch. The human mind is fragile; many feel exposed, uncertain, and small, as though sitting vulnerable in the dirt. I understand that challenging entrenched financial powers can seem perilous, as if it might destabilise one's own modest security, upsetting their small apple cart. People are conditioned to equate any call for economic fairness with dangerous ideology, labelled as un-American, a "Commie".

But this conditioning is precisely the problem. The influencers selling you a pipe dream of limitless wealth, the corporate media serving the interests of the already powerful, the YouTube gurus preaching the gospel of unchecked capitalism, the cult of Babylonian bling peddling fool's gold to the youth, these are not sources of wisdom. They are mechanisms of servitude, control and illusion. The reality is simpler: wealth persists not through some divine right, but through a system designed to protect it beyond gluttony. And what's required is not revolution, but reason, an intelligent recalibration of the system to ensure fairness, not just for the few, but for all, including those with wealth. Tell me, is Warren a fool? Watch, listen and learn, seek out the truth; Warren Buffett: *No one would owe 'a dime' of federal taxes if other companies paid their fair share.*

Mr. W.F. Burr for POTUS 2028, first executive order: Establish the Department of Anti-BS. And I'm not even kidding. Keeping it real is what matters. Unwavering honesty. Uncomfortable truths. The kind that requires strength to name out loud. I don't lean left or right, but Bernie Sanders' perception is sharp, his ethical perspective isn't some wild miscalculation. This is exactly why corporate media and big-money interests have always seen him as a problem. Because when someone tells the truth too clearly, they become inconvenient. Especially when they're not selling a side, just naming the pattern, and you don't need to be a revolutionary to recognise a rigged game.

But if things spiral out of control, democracy's strength lies in course correction, not blind loyalty to political tribes. When systems fail the people, the people can unite firmly yet peacefully. The real question is: Do they remember?

Meanwhile, European democracy drifts like a ship with a rusting hull, still afloat, still moving, but in need of urgent maintenance to ensure global stability. The far-right and the wealthy elite aren't just watching; they're circling with intent, casting bait, stirring division, and deploying well-orchestrated tactics to destabilise and manipulate.

Governments should be accountable to their citizens, not just to the highest bidder. Democracy, for all its imperfections, remains the best mechanism we have for course correction. The alternative is far worse. Better to endure the piranhas in the middle ground pool than to drift into deeper waters, where elitist sharks feed with impunity.

The Real World

So, what do we do about the real world, the one that sustains us, that we depend on for every breath, every meal, every future?

Politics aside, the reality is that children are held hostage. The young ones, the old ones, and those not yet born. That is where my heart is. That is where my mind is. And I suspect that for any rational, well-grounded person, it is the same.

At this stage, climate change no longer appears to be a looming threat, it is an unfolding reality. The evidence is plain to see, not just in scientific models but in the world around us. Yet, somehow, a certain percentage of the population still believes that the overwhelming consensus of climate scientists is part of some elaborate hoax, perhaps masterminded by the solar panel cartels? Let us keep it real. If the choice was between erring on the side of caution for humanity's sake or gambling with the future based on ideological stubbornness, the rational path was always clear. And yet, here we are, on the verge of tipping points that cannot simply be "*untipped.*" Glaciers do not regrow overnight. Why roll the dice?

To be fair, the scientific community itself is not entirely blameless. The betrayal of public trust by industries hiding behind lab coats, the chemical industry, pharmaceutical corporations, and the revolving door of regulatory capture has done significant damage. As a result, some people have developed a reflexive scepticism towards all science. But conflating climate science with corporate deception is a category error of enormous consequence. The question remains: Who, exactly, is bankrolling climate scientists? The Lithium mafia? The wind turbine Illuminati?

One thing's certain, we're in for a ride. Buckle up. This is just the beginning. Best to prepare for the worst and hope for the best.

The Theatre of Righteousness

If you spend enough time around human beings, you learn a predictable truth: those who struggle most with their own internal issues often seek out the moral high ground as a stage for their frustrations. This tendency can do more harm than good, and it is on full display within certain radical environmentalist groups. Self-righteous theatrics, however well-intentioned, have often played directly into the hands of the very corporate interests they seek to oppose. The corporations love this. They love the chaos, the division, the way people turn on each other instead of them.

Martin Luther King Jr. understood this. He did not rely on moral grandstanding; he studied deeply, applied his intelligence and heart, and built a movement based on strength, compassion, and an unshakeable moral foundation. He united people towards a common goal rather than alienating potential allies with hostility. If we are serious about changing the trajectory of our planet, we should take note. He gave humanity the blueprint. The past is written. The future is still open. What matters is what we do with it now. We must stay rationally calm and use intelligence.

The Plastic Problem

Plastic pollution is no longer just a visible blight, it is embedded in our oceans, our soil, our food. It is in our blood, the air we breathe, and even the placentas of unborn babies. Some estimates suggest we are consuming the equivalent of a credit card's worth of plastic every week, but the true figures remain elusive. What is clear is that plastic has become part of our biological reality.

The health impacts of microplastics are just beginning to emerge, and what we're seeing so far is alarming. The sharp increase in stomach and colorectal cancers among young adults could be an early sign of something far more insidious. While diet and lifestyle have always been part of the story, the speed and timing of this surge suggest there is another force at play.

Processed foods have been a staple in supermarkets for decades, yet this rise in digestive cancers appears to be a relatively recent phenomenon. Microplastics, now omnipresent in our environment, woven into our everyday existence, may be a critical factor. They introduce synthetic compounds into our bodies that could disrupt cellular functions, trigger inflammation, and serve as vehicles for toxic chemicals. If this proves to be the case, we are only beginning to grasp the long-term repercussions of living in a world saturated by plastic. The real question is whether we will act on this knowledge before the full scale of the damage becomes undeniable, before we fully realise what it means to exist in a world drowning in its own refuse.

This is not just about saving the planet, the planet will survive, but about whether we do, whether we wake up in time, whether we learn to think beyond profit, beyond the byproducts of materialism, and beyond the illusions of progress. This should concern everyone, not just environmentalists.

Addressing this problem requires more than feel-good "green" marketing campaigns and performative eco-capitalism. What we need is real data, real research, and systemic action. A bottom-up, individualistic approach will not be enough, this demands top-down policy changes and a fundamental upgrade in our collective intelligence.

Because, in the end, the only truly sustainable solution to keeping the planet balanced is for human beings to think and act with greater clarity and wisdom.

Chapter 22
The Anatomy of a Broken Species

Let's set aside the term "*ego*" for a moment. This word, derived from the Latin for "I" or "me," has become so overused and misapplied that it often obscures more than it illuminates. While "ego" can technically refer to the self, in practice, it is frequently used to suggest selfishness or self-centeredness. However, the way it is commonly wielded, especially in spiritual or self-help contexts, can introduce a layer of confusion, as if "ego" were some kind of mystical antagonist lurking in the shadows of our psyche, a boogie man hiding under the bed of our consciousness.

The problem arises when imagination supplants reason. When imagination starts driving the car, and reason is shoved into the back seat, that's when things get strange. If we allow abstract concepts like "ego" to take on an almost mythological status, they can become a source of delusion, distorting our understanding of ourselves and the world. Rather than serving as a clear and precise term, "ego" risks becoming a kind of intellectual placeholder, a word that pretends to explain but, in reality, merely perpetuates confusion.

For the sake of clarity, we might do well to abandon this nebulous term altogether. So, let's bin it. Strip it down to the bones. Call it what it is: selfishness, plain and simple. And when we name it right, we take away its power. If we are discussing the sense of self, let's be precise. However, continuing to invoke the "ego" as if it were a coherent, useful concept is risking muddying the waters of rational inquiry.

There's no need to invoke Freud, Jung, or any deep psychological analysis to grasp the fundamental issue at the core of humanity. The truth about humanity isn't buried in some dark, hidden corner of the psyche. When we get lost in layers of complexity, we risk missing the obvious. We need to clear the clutter, quiet the mind, and strip away the overbuilt concepts that keep us chasing our own shadows. The simplest approach is often the most effective. While every person carries a unique identity and personal story, at our core, we are all products of an unbroken evolutionary chain stretching back billions of years.

Your sense of self, your personal identity, is just the surface layer of who you truly are. Beneath it lies the weight of deep time, the accumulated wisdom of life's adaptive journey through countless generations. You are part of something vast and ancient. Without this evolutionary history, neither you nor your individuality would exist.

The unconscious is the land we walk on but do not see: the majority of who we are operates beneath the surface of consciousness. The unconscious mind, a vast and opaque domain, governs much of our behaviour, influencing our thoughts, emotions, and actions without our direct awareness. Just above this lies the subconscious, a murky bridge between the depths of the unconscious. And finally, at the helm sits the conscious self, the driver, who often believes it is in control but, more often than not, is simply along for the ride.

This body, this mind, this vehicle, evolved over aeons. Imagine your mind as driving an evolved supercar, a powerful yet complex vehicle forged in the crucible of millions of years of evolution. Mastering such a vehicle requires more than just the skill of driving; it demands an intimate understanding of what lies under the bonnet. Without this understanding, a host of issues can arise, misfires of thought, breakdowns of behaviour, and collisions with reality itself.

When we examine human beings, it becomes clear that not all drivers are the same. Some are exceptional, navigating life's complexities with skill and grace. Others are merely good, competent, but not extraordinary. Many are average, getting by without significant incidents. Then there are those who are lost, driving aimlessly through life, spinning circles in empty car parks. Some drive recklessly, endangering themselves and others. There are the drunk drivers, intoxicated by delusion or addiction, veering into dangerous territories. And finally, there are those who are careering out of control, their judgment burned away, with no regard for limits, self-awareness, or the well-being of others.

From birth, we embark on a journey of growth, physically, emotionally and cognitively. Childhood is a stage marked by immature traits: irrational fears, selfish impulses, and a limited understanding of the world. These traits, while seemingly negative, serve a purpose. They are part of the mind's natural development, providing protection and a scaffold for growth. However, if these traits are not properly addressed, if a child is not nurtured, not connected to the natural world, not educated in ways that promote individual growth, then a critical juncture is reached. It is here that things can become complex, ungrounded, and internally distorted.

As physical maturity sets in and desires intensify, the stakes are raised. If the mind has not matured in tandem with the body, significant problems arise. The imagination, a remarkable and uniquely human asset, can become a double-edged sword. Those who cling to immature traits into adulthood often construct a "*false self*" to compensate for what they lack in self-awareness and self-understanding. This false self becomes a mask, a persona that conceals deep insecurities. It can manifest as detachment, recklessness, arrogance, vanity, shallowness, heartlessness, and more. The R-complex, the strong primal part of our brain, combined with an unmoored imagination, can overpower the conscious driver, leading to a state where the mind is weak and directionless. The reckless can become more reckless and highly destructive. The lost drift further. And the sickness spreads.

This, fundamentally, is a sickness within humanity. Many adults remain psychologically immature, their bodies fully grown but their minds underdeveloped. Such individuals are more susceptible to deception, manipulation, and self-delusion. They gravitate towards those who reinforce their weaknesses; the weak elevate weakness. The blind follow the blind, perpetuating a cycle of dysfunction. In this context, humility, a grounded, clear-eyed understanding of oneself, becomes exceedingly rare. Instead, the world is often governed by the unwell, by those who mistake their sickness for strength.

This is where I respectfully have to take the gloves off; we need to shelve all the textbook psychoanalysis and keep it real. No one, in many ways, analysed humanity in a so-called advanced civilisation more insightfully "overall" than Charles Bukowski, mainly in his poetry, as others with perceptive minds have acknowledged, as in their quotes: "*Bukowski ate coal and spat diamonds*," widely attributed to Tom Waits, and "*He brought everybody down to earth, even the angels*" -Leonard Cohen. Bukowski's work, shared widely on social media, continues to resonate with honest, unflinching depictions of humanity's harsh truths, not because it is all doom and gloom, but because it acknowledges the full spectrum of the human experience. He saw hope, glimpsed the greatness within humanity, and recognised the flickering light of brilliance and beauty even in the harshest of realities created by mankind.

We've all made mistakes. That's part of the deal, part of the learning curves that come with the naïveté of youth or the disinhibiting effects of alcohol. Human social structures, these constructed systems we live under, often exacerbate such errors rather than mitigating them. Many of our norms and institutions seem almost designed to entangle us in poor decisions, encouraging impulsive behaviour instead of fostering thoughtful reflection. But the crucial point is this: Mistakes don't define us. Learning does. Mistakes, in and of themselves, are not the main problem. What matters is how we respond to them. Guilt, when it lingers and festers, can become a weapon turned inward, dulling the senses. It is rarely productive. It can cloud reason, leading us into a spiral of self-reproach that serves no one. The rational approach is to learn from our mistakes, let them go, and move forward with greater clarity and resolution.

The reality, however, is that not everyone learns from their mistakes in a constructive way. Some refine their errors like a craft, honing their ability to evade accountability. These individuals, those who display a profound lack of remorse, are particularly dangerous because of their calm and confident demeanour. Unlike the majority, who might exhibit signs of guilt or discomfort when doing harm, these individuals display none. Their victims, therefore, often miss the usual cues and are blindsided by betrayal or manipulation. The skill with which some people conceal their narrow-minded, predatory behaviour should not be underestimated. A lack of self-respect often correlates with a profound total disregard for others, and this can be expertly masked behind a façade of normalcy or cloaked in charm, masters of the bait and switch.

There are also those who do experience guilt but respond to it maladaptively. As Martin Luther King Jr. insightfully observed, "*Often people respond to guilt by engaging more in the guilt-evoking act, in an attempt to drown the sense of guilt.*" This is not merely a philosophical observation; it is a psychological truth. Many individuals, particularly those trapped in cycles of addiction or abusive behaviour, perpetuate their own suffering in a misguided strategy to numb their inner turmoil, inevitably dragging others into their destructive patterns along the way.

Personally, I am well-equipped to navigate these dynamics. Through a deep understanding of my own mind and a well-honed intuition, I can generally read people with a high degree of accuracy. But not everyone has developed this skill set, and many good, well-intentioned individuals find themselves manipulated, mistreated, or humiliated without even understanding why. It is important to emphasise that this does not make them foolish. There are countless reasons why deception succeeds, reasons rooted in psychology, social conditioning, and the fact that manipulators have had lifetimes to perfect their craft.

However, there is no need for paranoia. The best defence is a calm and measured approach to life. Avoid making assumptions. Fact-check wherever possible. Take your time when making decisions. Protect yourself through reason and logic, not fear. By cultivating this mindset, you build a mental resilience that not only shields you from harm but also allows you to engage with the world with clarity and confidence, and with a heart that refuses to be poisoned.

The truth is, if thoughts were criminalised, we'd all be serving time. Our minds generate an array of impulses, many of which are involuntary. It's not the thoughts themselves that matter, but our ability to recognise them for what they are, transient mental events. Thoughts are not crimes. The crime is in the action. The crime is in the harm we bring to others.

I've widely observed that kind and mindful individuals often carry a disproportionate burden of guilt, as though they were responsible for the world's brokenness. These are people with healthy empathy who, ironically, tend to be far too hard on themselves than those who genuinely warrant self-reflection. To those with empathy, I say this: Don't burn yourself out in the name of being good. Avoid the pitfall of unchecked altruism. It's not merely what you do for others that defines your character. True greatness is not just about acts of kindness; it's about what you consciously refrain from doing to others, choosing not to harm, not to exploit. Maintaining balance is key. While helping others can be virtuous, one must also walk with reason and preserve personal equilibrium. Respect the life that the universe, through billions of years of evolution, has afforded you also.

High IQ vs Intelligence

A high IQ does not equate to true intelligence. If intelligence were merely a measure of cognitive ability, pattern recognition, memory, or problem-solving, then some of the most monstrous figures in history would qualify as its exemplars. Serial killers, war criminals, cult leaders, men in suits who sign death warrants with a steady hand, and despotic politicians have often demonstrated extraordinary cunning. Yet their actions betray a fundamental deficiency, one that suggests intelligence, in any meaningful sense, must encompass more than raw intellectual cleverness.

Consider Colonel Russell Williams, a decorated officer of the Royal Canadian Air Force. His crimes were executed with chilling precision. He meticulously planned his offences, evaded detection for years, and exploited his position of authority to facilitate his depravity. By conventional metrics of intelligence, strategic thinking, discipline, and technical aptitude, he excelled. But what kind of intelligence leads a man to use his mind as a tool for unimaginable harm? To call such a mind "*highly intelligent*" is to commit a profound category error. His IQ-rated intelligence, if we insist on using the term, was not an asset to humanity but a pathology.

Marilyn vos Savant, who once held the record for the highest measured IQ, pointed to this distinction when she wrote: "*To acquire knowledge, one must study; but to acquire wisdom, one must observe.*"

And that's the thing. Knowledge is easy. But wisdom? Wisdom, the ability to make sense of experience, to learn from observation, to internalise ethical principles, is categorically distinct from intelligence as we conventionally define it. A person may accumulate vast stores of knowledge and command the subtleties of logic, yet they still remain a danger to others. Intelligence without wisdom is, at best, amoral. At worst, it is the engine of extraordinary harm.

What we often misinterpret as intelligence, particularly in dangerous individuals, is, in fact, an imbalance masquerading as high IQ. Their cognitive abilities are not directed by a fully integrated mind but are instead at the service of unexamined impulses. The unchecked dominance of primal instincts, the reptilian drive for power, control, or sadistic pleasure, untethered from empathy or higher reasoning, results in a dangerous distortion.

If our species is to evolve in any meaningful way, we must redefine intelligence. True intelligence is not simply raw processing power but the ability to integrate knowledge with wisdom, morality, and self-awareness. Anything less is intellect running on instinct, an instrument of destruction rather than understanding.

In a broader sense, this sickness isn't new. It has existed for a long time. I want to use a metaphor to illustrate a kind of psychological detachment, a sickness within humanity that manifests on a large scale because we live in a time when it is everywhere, shaping the way people see, think, and act. And we need to recognise it for what it is. While this is my personal interpretation, I believe it captures something fundamental, an intuitive vision, beyond religion, that is both obvious and deeply metaphorical.

Consider Albrecht Dürer's artwork, Durer, apocalisse, 12 il mostro marino e la bestia. *The Beast from the Sea.* The 'sea' in mythology often represents the depth of the greater unconscious, a vast and shifting realm. The seven-headed monster in this piece can be understood as representing seven distinct yet fundamentally similar ways in which human beings can detach from true intelligence and moral grounding. Each head wears an off-centred crown, symbolising self-centredness, shallowness, and a detached, self-serving mentality, a reflection of how human beings twist themselves. Though they may appear different, they share the same underlying flaw, depicted in the same body: a disconnection from wisdom, individuality, and deeper intelligence. You see minds lost in their own illusions, circling the same patterns again and again, because they have no real grounding, individuality, no true connection to anything beyond themselves.

One of the more compelling e lements of the artwork is the figure, often described as a beast with lamb's horns. I would argue that this is not a beast at all, but rather a lion. The lion represents humanity's intelligence and strength when aligned with the heart, while the lamb signifies innocence and kindness. Together, they form a powerful symbol: the unity of strength and moral clarity emerging as one, fierce, but not *cruel*. A force that sees the sickness, that understands what it is, where it comes from, and how it manifests. And this combined force it is rising, emerging from beneath the weight of oppression, ready to confront the monsters of detachment. The spectators in the piece are complicit in the dominant forces, benefiting from them in ways that make them either oblivious or wilfully in denial. Some refuse to see, because to see would mean taking responsibility.

Other elements in the artwork, such as God depicted in the old way, as an elderly wise man surrounded by angels, are simply epochal representations of higher intelligence. The concept of "God" is not something I will define, though I find the Native American notion of the Great Spirit, a profound mystery rather than a rigid doctrine, closer to the mark. Whether such an intelligence exists is a matter for science to explore, but what remains clear is that Dürer's imagery captures something real: a broad state of detachment that can exist in individuals of both high and low intelligence, a sickness that manifests in ways consistently destructive to human progress.

But here is what matters: the sickness does not care how intelligent you are. It does not care if you are rich or poor. It infects those who allow themselves to become lost in it.

If you observe closely, you may notice a pattern. Despite appearing outwardly distinct, certain people share an underlying similarity, a hollowness, overgrown immaturity, and a lack of genuine individuality. You see it in certain groups, though seemingly separate, operating with the same destructive intent, bound together by a shared psychological flaw. I will leave it to others to reach their own conclusions, but from experience, true intelligence, true strength, reveals itself in a different way.

Having worked in a wide range of jobs and put myself through education from the age of eighteen, I have encountered both exceptionally strong individuals with compassionate minds and those who were less outwardly strong yet deeply genuine. What distinguished them was not their IQ or financial status, but their moral fibre. No matter the circumstances, they would not knowingly harm others for their own gain. The ones who have minds that see, hearts that feel, no matter how much power or wealth they acquire, will not betray their principles. Their individuality remains intact.

The universe is always searching for balance. Perfection is an illusion, but balance, that is something else. There comes a point at which imbalance becomes so extreme that it is no longer just deviation, but something deeply harmful to society and humanity's evolution. What we call "*evil*" can now be understood not as a mystical force, but as a real sickness, one that manifests in varying degrees yet remains present and undeniable.

The Education System: A Failure of Priorities

Let's take a moment to examine our education system with fresh eyes. We see a structure designed not to cultivate wisdom, resilience, or self-sufficiency but to condition individuals for an economy that demands compliance. A system that produces workers, not deep thinkers. Consumers, not creators. It teaches the young to earn, but not to live. From childhood, we are trained to follow a linear trajectory, one that prioritises employability over self-sufficiency. It teaches dependency, not connection. The fundamental question we must ask is this: What does it mean to be educated?

A truly well-rounded education should equip people with life skills, the tools to navigate life rationally and sustainably. It should place as much emphasis on essential survival skills as it does on abstract academic pursuits. Yet how many young adults graduate without understanding how to grow food, cook a meal from raw ingredients, or maintain their own living spaces? The consequences of this neglect are everywhere, rising rates of chronic illness, dependence on industrial food systems, and an increasing psychological disconnect from the natural world.

We cannot call ourselves an enlightened and educated society when the most basic requirements for human health and independence are treated as niche interests rather than fundamental components of education. We face a crisis not just of knowledge, but of values. In many cases, children must educate their own parents, bringing home insights that, if applied, could improve the health and well-being of entire families. This is particularly urgent when we consider how conditions like diabetes, once rare, are now widespread, a predictable result of an economic system that prioritises profit over well-being.

Even from a strictly pragmatic perspective, life skills should not be seen as optional. Self-sufficiency is not a fringe concern due to natural disasters, economic downturns, or unforeseen disruptions; it is a form of resilience and self-worth. And yet, our educational institutions systematically avoid fostering this kind of empowerment. This raises an obvious and uncomfortable question: Who benefits from a population that is dependent, uninformed, and disconnected from the mechanics of its own survival?

The answer is as straightforward as it is unsettling. A population that does not know how to provide for itself must rely on external systems, corporations that supply its food, industries that build its homes, and financial institutions that keep it locked in cycles of debt. A truly empowered citizenry is far less profitable. When people understand how to construct and repair their own living spaces, grow their own food, and make informed choices about their health, they become less dependent on industries that thrive on their ignorance.

It's time to wake up. Time to educate the education system. This is not a radical proposal; it is simply a rational one. If education is meant to prepare individuals for life, then it should reflect the realities of living. Hands-on experience with sustainable building techniques should be as integral to a curriculum as mathematics. The construction industry's profit-driven model, where buildings are designed for planned obsolescence, has conditioned us to accept short-term solutions over intelligent, long-lasting design. But there are far superior ways to build that are ecologically sound, financially sustainable, and structurally resilient. The fact that these approaches remain outside the mainstream curriculum tells us everything we need to know about where the true priorities lie.

Consider what could be achieved if communities, especially those facing economic hardship, were educated in practical, self-sustaining skills. Imagine a generation that does not simply consume, but creates; that does not rely entirely on external systems, but understands how to engage with and improve them. This is not about rejecting modern civilisation; it is about making it more intelligent. Because, in the end, education should not be about keeping the machine running. It should be about freedom.

A System That Fails Its Own Students

It is no wonder that so many young people disengage from education, not because they lack intelligence, but because they sense, often unconsciously, that the system is not designed for their benefit. They are trained to function as cogs in an economic machine, to exchange their time and energy for a paycheck, and to take on debt in pursuit of credentials that may or may not lead to financial stability. They are taught to see learning as a means to an end rather than an inherently valuable pursuit.

The predictable result is a generation that seeks escape into screens, dumbed out by smartphones, lost in sugar-coated entertainment, distracted by temporary reprieves from a world that feels fundamentally misaligned with their deeper needs. The education system, as it stands, does not cultivate intelligence in the broadest sense; it narrows it, channels it, and, in many cases, stifles it.

If we are to foster a society in which intelligence is used for the betterment of all, rather than simply for the pursuit of wealth or status, then we must rethink the foundations of education itself. There is nothing inherently wrong with academic ambition, nor with the pursuit of success in traditional fields. But these should not be the only paths presented as valuable. A system that serves only the few while neglecting the many is, by definition, a broken system.

Relationships and Raising Children

A relationship built on anything but truth is already crumbling. If it is not friendship, if it is not teamwork, it's on shaky ground, whether romantic, familial, or otherwise. Relationships must be built on a foundation of mutual respect and cooperation. And being a parent, that is not just a right, it is a responsibility. A duty to do right. Children may be young, but they are not empty vessels to be programmed; they are evolving minds, learning more from what they see than from what they are told.

One of the most damaging things we can do is underestimate the intelligence of the next generation. Children are naturally attuned to the world around them; they sense hypocrisy, recognise contradictions, and instinctively rebel against systems that feel incoherent. If we wish to cultivate a generation that is not only intelligent but also wise, we must examine the environments in which they are raised.

A spoiled child, who is taught, through action, that the world exists to serve them, will carry that expectation into adulthood, often with disastrous results. Screens may entertain them, but they will not raise them. Encourage them to connect with the natural world and engage in real experiences. Give them constructive encouragement, or they may become easily discouraged later in life. Teach them to organise, to shape their lives with intention, because life itself is a force of order moving through the chaos. Guide them with kindness, for kindness is a shield against the sickness of the mind-against fear, against self-doubt, against the loneliness that creeps in when the world feels hollow. It fosters resilience, and critical thinking will grow into an adult capable of navigating life with clarity and confidence. Show them self-respect by respecting yourself. Be their best friend, for in the end, they will walk the world with the lessons you left in their hearts.

And remind them, always, that they are not small. They are not weak. They are made of the same power as the stars. And the universe loves them, deeply...Just as it loves you.

Chapter 23
Contact without Contact

Reluctantly, I decided to include this. The last thing I want is to discredit myself or be aligned with conspiracy theorists, lumped in with the tinfoil hat crowd. But ultimately, this is a matter of human experience, one that, if real, has profound implications in a humanitarian sense. If not, then at the very least, it presents an interesting psychological or perceptual phenomenon worth considering.

For the record, scepticism is necessary. Be sceptical. Dismiss this entirely if you like, or simply set it aside as something to consider or revisit later. But what I saw, what I lived, it was real, and I will do my best to describe it in a way that remains grounded. I will not go into detail about how I achieved this, other than to say it was my understanding of dimension. No drugs of any kind were involved.

What I encountered was not of human origin; that much was clear. And yet, I came across no extraterrestrial beings, no little green men or cosmic overseers. The experience was more quantum in nature, as if I had gained access to an underlying process, one seemingly designed not to control but to assist intelligent life. It was as if I had stepped into something that functioned with a kind of profound coherence. It wasn't built to control. It was built to guide.

This was not a crude technological mechanism but a system that honoured life as the pinnacle of the universe. I wasn't simply imagining it. I was there, fully aware, while still present on Earth. The process revealed itself to me like a thin, translucent film that I was somehow within. I could see it, navigate it, and yet still perceive our reality with total clarity.

For two days, I moved within it. I saw its patterns, its intelligence. It wasn't alien, yet it wasn't human. It was like a bridge, an interface. I did my best to understand its structure, though for reasons unknown, I was confined to a particular region of the system. It was an astonishingly precise experience, geometric yet fluid, structured yet artistic. The mechanism itself was not just elegant, but deeply advanced, leaving me in awe of its complexity.

I stood upon two highly advanced rotating pedestals, one beneath each foot, which interacted with me in a way I still struggle to articulate. They were larger than they appeared, containing a sophistication beyond anything I had ever encountered. And from there, I was elevated into the process.

Within this structure, the thin film led to what felt like a larger chamber, though this wasn't a room in any traditional sense. It existed in a space more like a void, without walls, yet possessing shape and dimension. And there, in the heart of it, stood two figures.

Monolithic. Human, shaped, like recesses, like grand statues or sentinels. This may seem slightly absurd, but their form closely resembled the Oscar statuette in style, simplicity and contours. Yet the Oscar itself bears a striking resemblance to Ptah, the ancient Egyptian god of craftsmen, to such an extent that I strongly suspect it may have been modelled on, or at least inspired by, the mythical god. These figures lacked intricate features, yet their presence conveyed an unmistakable calm authority.

One white, the other black, perhaps symbolising an inner and outer aspect of the mind, a duality within human cognition itself. These figures were motionless, yet they were integral to the system. Their purpose wasn't clear, but their presence was undeniable. The statues remained fixed, yet they were clearly essential to the broader process. Their significance wasn't in movement but in their relationship to an intricate system, one that seemed to interface directly with the nervous system itself.

It was then that I recognised something deeply familiar. I had encountered this phenomenon before, albeit in fleeting moments throughout my life.

This process, at its core, involved a weaving structure, one that wrapped around and permeated the nervous system in a way that was both precise and deeply affecting. There were two distinct aspects to it.

The first was soft, a white weave, something internal, something that soothed, that comforted from within. The second, an outer layer, provided a stabilising sense of wholeness, as if reinforcing the structure of the self from without. To witness this in motion was extraordinary. The weave moved with an intelligence of its own, reacting, adjusting, and enveloping.

I understood, in part, the state of being required to induce each weave, and somehow, this process was able to read my nervous system in real time, responding to my condition. There was also a thin, barely-there, membranous film surrounding the experience; its nature became clearer when I turned my focus to the natural world, when I "aligned" myself with it. As I did, the film erupted into breathtaking, shifting nebular hues, as though revealing its true composition.

In that moment, I became acutely aware that I was inside some kind of tunnel-like structure, perhaps a wormhole or a filament extending, a bridge between realities.

The purpose of this process was singular: to realign human beings with nature. It was not about transcendence in the mystical sense, nor about escape, but rather about returning to a more natural, grounded state, to reconnect with the Earth, particularly after illness, trauma, or some form of nervous system imbalance.

I came to realise that there was yet another layer, a golden weave, distinct from the others. This one was not present in the process I observed, yet I knew it existed because I had encountered it before, though not in the same way. My best explanation for its absence was that it becomes imperceptible once a person has fully recovered or realigned. Its function, it seems, is to reinforce groundedness, acting as a final bridge between the individual and the natural order.

I am certain that many have unknowingly experienced these processes, perhaps in childhood, during moments of deep restoration, or in times of recovery. These sensations, often dismissed as normal, now appeared to be part of a far deeper system, one that had always been there, operating quietly beneath the surface of human awareness.

This, inevitably, brings us to mythology, not in the sense of arbitrary stories, but as historical echoes of this very process. The imagery of the Four Horsemen, for instance, could very well correspond to the weaving mechanisms I encountered. The rocking sensation of the experience, the interplay of forces, and even the colours align with certain mythological traditions.

The legend of the White Buffalo Woman, too, contains four colours and a motion that bear similarities to what I observed. These correlations suggest not that these myths are literal descriptions, but that they may be ancient interpretations of a process that has existed for eons, a process that some have, on rare occasions, directly experienced.

The Egyptian Correlation

If there is a historical reference point that aligns with this process, ancient Egyptian symbolism provides one of the most compelling examples. Tutankhamun's mask, for instance, offers a striking visual, some deeper parallel to the weaving mechanism I observed, though the colours may not be an exact match. What stands out most, however, is the central point on the forehead. It appears to correspond not to a mere decorative feature, but to an essential function of the process itself, one that seemingly splits awareness in two as the weaving unfolds. A place where perception divides, where awareness turns in on itself. The reason for this remains elusive, but it is evident that Egyptian iconography depicted more than mere artistic expression. The patterns, the geometry, the symmetry, they all speak. It is the language of something older than kings and empires.

The structural and artistic choices of the Egyptians suggest an intimate knowledge of this phenomenon. The architecture, the precise geometries, the profound artistry, they all point towards something deeper. My conclusion is that this knowledge was acquired through direct experience, perhaps by individuals with extraordinary cognitive abilities, such as savants, or by high priests who accessed altered states of consciousness through specific means.

It is conceivable that these individuals employed psychoactive substances to open the senses, to break through the walls of perception. Peganum harmala, also known as Syrian rue, is one such candidate. Its seeds can induce vivid, dreamlike visions, altering perception in ways that might have allowed for a more direct observation of this process. Similarly, the Egyptian blue lotus, a mild sedative known to produce euphoria, may have played a role. A combination of these substances, properly dosed and administered, could have enabled these early seekers to witness the underlying structure of reality in ways that remain inaccessible to most.

This perspective also casts new light on the Egyptians' seemingly anomalous grasp of geometry and engineering, an understanding that appeared to emerge with extraordinary sophistication some 4,000 years ago. It suggests that what we perceive as a sudden leap in technological and architectural capability may, in fact, have been the by-product of insights gleaned from this process.

The pyramids themselves, then, may not have been designed purely as tombs or monuments, but as echoes of something they glimpsed, an attempt to replicate or harness the fundamental forces they observed. They may have sought to construct something analogous to the system they perceived, perhaps a structure that, in their understanding, functioned as an energy source or a stabilising mechanism for these processes.

That said, while the pyramids remain a feat of unparalleled ingenuity, their construction is not beyond human capability. We do not need myths of ancient astronauts. The logistical challenge was immense, but not insurmountable. With an estimated workforce of 20,000 labourers over two decades, such an undertaking becomes conceivable. Ingenious techniques, such as those demonstrated in the video - *Ancient Stonehenge Technology* Revealed by Wally Wallington, illustrate how simple mechanical principles, properly applied, can achieve what might otherwise seem impossible.

Thus, I find no compelling reason to invoke extraterrestrial intervention. The idea that non-human intelligence played a direct role in human development is, in my view, unnecessary. What the Egyptians achieved was not the result of outside interference, but of human ingenuity, vision, and an unparalleled ability to translate extraordinary insights into lasting physical structures.

A Fundamental Problem

There is a problem.

Within the process, I noticed a clear and troubling obstruction, one that stood in stark contrast to the natural flow of what was unfolding. The right side was severely clogged. A blockage, thick, tangled, on the right side. I could see it. I am certain this has to do with humanity's widespread failure to engage the rational faculties of the mind. Many people, whether they realise it or not, seem to be caught within the process itself, stuck in a state of cognitive and perceptual misalignment.

I remained in this state for too long. It drained my nervous system beyond exhaustion, well past any ordinary depletion threshold. I was trapped in it. For five days. I could not fully perceive what I needed to do to be released, though I could still feel it, an unrelenting presence. The only way out was through natural neurological recovery, yet the paradox is that when one is caught in it, such recovery becomes exceedingly difficult to achieve.

This experience led to a number of insights about human cognition and health-insights significant enough that I feel compelled to include them in this book. That is why I am writing this. That is why it matters.

The intensity of being trapped in the process was staggering, akin to being caught in a relentless, grinding mechanism. It was not just psychological or neurological but physical in an undeniable way. Intense tinnitus and a metallic resonance accompanied the experience, as if I could hear the process itself moving. Alongside this, my body rocked with it, as though caught in an inescapable current, accompanied by an extreme head-to-toe sensation of pins and needles. The weaving motion would not stop; it's a relentless, perpetual process.

At a certain point, the rotations themselves began to damage my nervous system. This brings me to a few conditions that I believe are closely related to this phenomenon. Take fibromyalgia, for example. If you examine images of its mapped pain points or have experienced it yourself, you may recognise a similar underlying mechanism at play. The sensation is not superficial; it is something deeper, something occurring beneath the skin. Its intensity is difficult to overstate.

During my recovery, I also experienced migraines, not the standard headaches most people associate with the term, but something more intricate. Bright, geometric shapes and lines, kaleidoscopic distortions, zigzagging patterns. This is a commonly reported experience among migraine sufferers, and it seems to align with aspects of the process itself. However, I was able to mitigate some of these effects through specific techniques, methods that engage with dimensional awareness and the natural world in a way that offsets neurological overload.

Stepping beyond religious interpretations for a moment, some ancient descriptions of visionary experiences seem to mirror this process in striking ways. Take Ezekiel's Wheel, for instance, its symbolic depiction may, in some way, represent the same fundamental structure. Even Carl Sagan's Contact offers a compelling visual parallel. The massive rotating machine encircling the pod bears an uncanny resemblance to this process, albeit on a much larger scale than the one surrounding me in that moment.

This also sheds light on some of the more bizarre and frequently ridiculed aspects of so-called alien abduction experiences, particularly the reports of "anal probes." At first glance, such claims seem absurd, an invitation to public humiliation. But that is precisely what makes them worth considering. Who would willingly claim they were taken, examined, violated, knowing full well they'd be mocked, unless they were truly convinced of what they had experienced?

The reality, however, is likely far less theatrical. A discomfort so deep that the mind scrambles for an explanation. When caught in this process, one can experience an intense strain in the pelvic region, radiating up the spine. It's a function of opposing forces, an upward pressure from below met with a downward compression from above.

A profoundly uncomfortable phenomenon. In the disorientation of altered states, it's easy to see why some interpret it as an external event, something "done" to them.

Similarly, sleep paralysis, so often linked to alien encounters or shadowy presences, may be another byproduct of this process. If the nervous system is drained of a particular type of energy, the body can enter a locked-in state, where consciousness remains active while motor control is temporarily disabled. This is not abduction; it's a well-documented neurological event. And for those who have suffered through it, you are not mad.

But here's an important warning: much of what is packaged as "spiritual awakening" or "chakra activation" appears, in my view, to be a misinterpretation of this process. Certain sections align with those mapped-out energy points, but that doesn't validate the mythology. The so-called "kundalini awakening", with its spontaneous spasms, tremors, energy jolts, and manic highs, is not enlightenment. It's nervous system instability. This is why individuals withdrawing from drugs report similar symptoms: depletion of neurochemical reserves leading to electrical misfiring throughout the body.

The lesson is simple: don't chase the high-end states. Reality, coherence, clarity, and grounding is the safest and most sustainable path forward.

And yet, I'm not comfortable with what this suggests. I don't like this process. Not at all. It feels like an intervention, something imposed rather than emergent. But if it is, why the flaw? If intelligent beings engineered it, surely they had good intentions. And yet, if humans suffer because of it, they either failed to anticipate the consequences or were never as intelligent as we might assume, even though technologically advanced. Perhaps they are extinct.

Or this process was never meant for us, and human intelligence grants us access to something unintended. This ties into my hypothesis on our *primordial genetic code*, that our biological evolution follows deeper principles dictated by gravity and fundamental laws of the universe. If true, it raises profound questions about the origins of this process and how it intersects with our reality. But the means to resolve those questions remain far off.

For now, there is *no* off-switch. There is no clear mechanism to halt or control what's unfolding, not without a far deeper understanding. Perhaps one day, quantum computing will offer insights, but as things currently stand, it's overhyped and inelegant: clunky, even. The depth and plasticity of the human mind still dwarf all computational technology. Next to life itself, it remains crude.

That said, AI and advanced computing could assist humanity. They may become valuable allies. But, like any tool, their impact will be entirely dependent on the wisdom, or folly, of those who wield them. Einstein understood this well: technology in the hands of fools is a recipe for catastrophe.

So, scepticism isn't just appropriate, it's vital. Many will scoff, and that's fine. This chapter departs from the confines of conventional discourse. I include it not to convince the sceptics, but for those who are suffering, those in need of frameworks to restore neurological balance. The "Unlock method," as outlined here, is offered in that spirit. It may help mitigate the destabilising effects and provide a path back to coherence.

Even though, statistically speaking, life beyond our solar system is not only possible but overwhelmingly likely, nothing quite prepares you for the actual shock of experiencing that reality. It was existentially jarring. And yet, my concern remains here, grounded in the well-being of humanity. The more neurologically stable and cognitively clear we are as individuals, the more coherent, compassionate, and resilient we become as a civilisation.

Finally, I encourage a healthy degree of doubt, especially regarding extraordinary claims. There are those who exploit human vulnerability, parading pseudo-spiritual insights for personal gain. Sadly, that domain is saturated with deception. Figures like James Randi did essential work in exposing this, lifting the veil, revealing the sleight of hand, and reminding us that what ultimately matters is evidence. Hard data. Verifiable truth. And I hold myself to the same standard. What I've shared here is speculative. But I plan to follow wherever the facts lead, bringing evidence with me, not ideology.

Chapter 24
The Shape of What Remains

Time had slipped through my fingers, dissolving into conversations and discoveries, each moment folding into the next. I hadn't realised how quickly it had gone. My retreat had come to an abrupt end, leaving no room for ferry rides or family visits. London was waiting. The museum was waiting. The unrelenting pulse of the city, waiting. And I had no time to spare.

An early night was essential. Tomorrow morning, I'd be on a direct path, straight to the airport, Puddles curled in his carry pack, a flight to Heathrow, then Brit's.

TR had sent a brief outline of what he had managed to include in the book, with the rest to follow in a week or so, something I'd place after the Theory section. As expected, his insights were nothing short of striking: sharp, rational, unwaveringly precise. His emails held a cadence, each subject distilled with clarity, pulling complex ideas into coherence. They sparked something, a quiet, simmering energy, an unfolding of questions I hadn't even known to ask.

I had writing ahead of me. A great deal of it. The journey that had led me here, the moments, the encounters, the shifting of everything I thought I understood, needed to be recorded. The sheer weight of it settled into me, not as a burden, but as a quiet force of purpose. Love was woven through it, through me, steadying my mind, aligning my body. It deepened my understanding, folding the external world inward, shifting my perception into something more fluid, more whole.

Existence simply made sense. Profound. Clear. Soothing.

TR had described it once, how it was almost a cosmic fluid syllable, something beyond the grasp of language, beyond human comprehension, a resonance, an awareness, like music from an ancient, unreachable place.

I called Mum to let her know I'd be heading straight for the airport in the morning, there simply wasn't time. The flight would be tight enough without testing speed limits. She was a little disappointed but understanding.

I wasn't surprised when she began telling me about her dreams, vivid, shimmering, filled with wonder. She said she felt like a child again, as if she had rediscovered something whole and untouched, something once lost to a world that had long since tilted off balance.

Even Dad seemed more lucid, more himself, spending time in the garden, quietly tending to things as he once had.

Before bed, I wrote to TR, asking if there was some correlation, if this, too, was part of something more, or if it was simply chance.

And I had to include one final thought. There was no doubt in my mind what the book should be called. I didn't know if TR would agree, but after everything, the words he had shared, the Theory itself, the strange and inexorable sense of knowing, there was only one title that felt right.

Meeting the Messiah: The Theory of Everything & The Holy Grail.

And then, as if right on cue, his reply arrived:

Hi D,

Glad to hear your parents are doing well. It doesn't surprise me. Love has a cascading effect; the current flows. It tends to work that way when one is relatively centred and grounded, as yourself. That said, I suspect the effects on your father, given his condition, may be temporary. Parkinson's, as you know, is a complex neurodegenerative disorder, and while I haven't conducted direct trials in this area, I suspect what is happening is akin to the benefits seen in deep music therapy: a transient but meaningful shift in neurological state.

All the same, anyone claiming to be a mystical "healer" is engaging in deception and duplicity, selling illusions. What you and your family are experiencing isn't a miraculous intervention; it is the natural response of the mind to certain conditions, or more precisely, to love itself, moving through, aligning things, even if only for a while. I can facilitate alignment differently, but the fundamental mechanism remains the same. And no matter how deep or profound the experience, we need the weight of truth. While anecdotal experiences are valuable, they do not replace rigorous study. I aim to help, as serious research is required to understand these effects more fully, perhaps offering insights into consciousness, neuroplasticity, and even new avenues for treating neurological conditions.

As for other diseases, for suffering. I may be able to provide some degree of comfort, a little ease in the storm, perhaps a reduction in pain or anxiety, but beyond that, I hold no illusions. Life unfolds as it does, and medical intervention must remain rooted in evidence and precision.

The book title? We use it. The weight of it. I'll carry that burden. It has to land somewhere, and it has to lead somewhere. People need to wake up; they need to understand, as Monty Python put it in Life of Brian, "*You don't need to follow me, you don't need to follow anybody. You've got to think for yourselves. You are all individuals.*" Einstein echoed the same sentiment: "*Everyone should be respected as an individual, but no one idolised.*" Both strike the same rational chord.

If I can play a part in tearing down the walls, cutting through the noise, stripping away the illusions, then maybe, just maybe, we can help humanity rise beyond the pull of tribalism, the hunger for more, the quiet comfort of not knowing.

Because true leadership is not about gathering followers. It's about lighting a spark with wisdom. Giving people the power to think for themselves. To stand on one's own feet. When enough minds wake up and enough people start seeing clearly, something shifts.

That's when true freedom rises, and tyranny fractures, then begins to crumble.

Because in the end, a society is only as strong as the individuals who shape it. And together, under the banner of greater truth, we take the first great step forward, led not by fear, not by blind faith, but by reason.

The Grail, as a metaphor, recurs throughout this book, representing more than one path to deeper understanding.

This book, it's about healing the mind, breaking the cycle, opening doors that have been welded shut for too long. It is about contributing to humanity's well-being and opening vast new frontiers of discovery.

Safe travels back to London.

Truth and Love,

TR

TR, passionate, unwavering, driven on by something greater than himself. A mind that saw beyond the surface, a heart that carried the weight of truth. I knew love had a ripple effect, something deeper than mere coincidence. What was happening with Mum and Dad wasn't random. Not after everything I'd experienced this past week, not after the conversations that had shifted something in me. TR was humble, but his message? Unshakable. And I felt it, deep in my core, something no words could ever fully contain, awakened.

Idolisation, this waiting, this blind hope for some external saviour to set things right, was the oldest deception of all. A fool's game. We are each responsible for the intelligence we've been given. Surrender that, and humanity remains shackled, circling the drain of its own making, doomed to repeat the same ruinous cycle.

The title wasn't just an ending. It was a reckoning. A closing of one chapter so we could step forward, not as blind followers, but as something stronger. As individuals, aligned where it mattered most. The world had been drained dry by grifters for centuries, their words spun into silken lies, wrapped in the illusion of hope, veiling the selfish gain of the few. And the masses? They had fallen deeper, tangled in a web of deception, their true birthright stolen from beneath them.

I awoke the next day and made my flight in a blur of chaotic navigation, a full-scale traveller's triathlon toward my destination. By the time I finally tumbled through Brit's flat door, I was wrecked. Puddles wriggled free from his carry pack, triumphant.

I then made myself a strong latte, slumped onto the couch, and let out a sigh that could have come from a horse clearing the Grand National finish line.

Meanwhile, my pint-sized jockey, whom I'd faithfully carried to victory, strutted straight to the kitchen, rubbing himself against the bench corners, sniffing for a snack.

'No chance, mister,' I muttered, watching him angle an expectant look my way.

Mum had plumped him up enough. I wasn't about to be an accomplice.

A single, pleading meow.

I narrowed my eyes. 'Water. That's the menu on offer.'

A flick of the tail, a slow blink. The standoff had begun.

After fifteen minutes sprawled on the couch, gathering my energy that had scattered like dry leaves in the wind, I pulled myself upright. The weight of travel still clung to me, but I dragged my suitcase into the room, methodically unzipping, unpacking, and returning things to their rightful places as if restoring some small order to the universe. Then, the laptop opened with a familiar flick of the fingers, the screen glowing in quiet invitation. Puddles, meanwhile, wove his way through the flat, reacquainting himself with Brit's space, padding lightly, tail flicking, pausing now and then to sniff a pot plant, pawing at unfamiliar leaves, as if noting the subtle changes since his last visit.

It had been over a week since I had visited the Forum, a week absorbed in TR's world, in the cottage's quiet magic, in the unravelling of something both intimate and immense. The screen refreshed; the threads scrolled before me. A noticeable shift, a flood of posts in the dream-vision section, people writing of changes, subtle yet profound, an unmistakable current running through their words. They spoke of grounding and clarity, of an uplift in their well-being, of dreams feathered with insight, patterns emerging, something awakening.

And then, the Egyptian themes. A thread among threads, gleaming like a buried relic, strikingly aligned with TR's reflections, as if the knowledge had begun to echo, ripple outward, touching minds in ways unspoken. It was impossible not to be drawn in.

I sent a quiet message to Anna, cautious yet open, telling her only that my time with TR had been deeply moving, that it had kept me from the Forum, but that I was here now, returning, adjusting. I did not expect a swift reply, it was past one in the morning where she was, but the screen blinked, the message arrived:

Hi there, sister. I stayed up late, sensing I might hear from you. I had many strong visions of you and TR sitting beneath the great Tree of Life. I know of the book, the book to begin the Great Awakening. He reached to the stars and handed it to you, its pages unfolding like wings, carrying more than words. So many visions, D. They wrapped around me like clouds.

Blessings,

Anna

I sat back, pulse quickened. The book. She had seen the book. Surprise flickered through me, though not disbelief-no, not entirely. Hadn't Mum, too, spoken of the Great Tree, of clouds gathering, circling? The Nostradamus Vaticinia Code, plate 67, the Book of Life, not just a book, but something fluid, something shifting beyond language, beyond the limits of expression. The theory, yes, but also the knowing. The weight of what could not be explained but only felt.

The sound of Brit clattering into the flat broke the moment, boots scuffing, a rustle of gardening gear, her voice lilting as she greeted Puddles, asking him where I had disappeared to. The world folded back into the tangible, the immediate.

I typed quickly:

Yes, Anna, it was a transformative time, to say the least, and your vision was in many ways precise. Just back from Ireland. My friend has just come home, so I'll talk to you again soon once I've settled in.

Love, D

Send. The message dissolved into the ether. I closed the laptop, exhaling, feeling the weight of it all settling into place.

I wandered into the kitchen and straight into a big, dirty hug, literally,the kind only Brit could deliver after a day spent frolicking about in the dirt, knee-deep in the business of making the world bloom. Bone-crushing, unapologetic, the scent of earth and toil clinging to her, the residue of grafting woven into the fabric of her clothes. There were specks of compost trailing from her sleeves, a dusting of something distinctly cow-manure, like settling over her like a badge of honour. She pulled back, grinning as she flicked a few stray bits of straw from my shoulder.

'So, you made it back to the big smoke, then. Obviously.'

She was already half out of her bib and braces, peeling them off in practised movements, her boots being nudged off one by one with a satisfying thunk against the floor.

'Sorry again for not picking you up from the airport. The gardening job today, absolute nightmare, had to be done, no way around it.' She huffed, pushing a hand through her hair, a dusting of her long day floating away in the motion. 'I would've sent Paul, but the thought of you and Puddles strapped to the back of his electric bike, whizzing along the M4 at breakneck speed, well, not exactly what I'd call safe, would you?'

I snorted, the image forming instantly. Puddles, flattened by the wind, ears pinned back, his little paws clutching desperately as Paul weaved through traffic like some unhinged road warrior.

'Oh, I don't know,' I mused. 'Puddles might've enjoyed the thrill. And I would have donned a spunky pair of antique aviation goggles for the occasion.'

A loud thud behind us. Puddles, unimpressed by the very notion, had leapt onto the counter, fixing Brit with his signature love-me eyes. Brit chuckled, scratching his bonce as she busied herself with the sacred art of coffee-making.

'Now, D, you're a bit of a dream junkie, so I have a peculiar event to share,' she said as we drifted toward the balcony, coffee cups warm in our hands. 'Had this dream, not wanting to sound like my crystal-rubbing sister, but I was doing breaststroke in some kind of celestial pool of love.'

She plonked herself down in the cane deck chair, watching me, watching the sky, watching her own words linger in the air before continuing.

'Weird thing is, though. Paul had the same type of dream. Same night.'

I tried not to smile. I had a solid hunch as to why. I just raised an eyebrow and took a slow sip of coffee.

'Interesting, Brit. Maybe you both had too much cheese before bed. Pizza, perhaps?'

She frowned, sniffing at a climbing rose in bloom. 'Hmm. No.'

She paused, a flicker of something crossing her face before she continued, nonchalant but pointed.

'Anyway, Paul and I had a big love chocolate session the next day,' she laughed, shaking her head, even had a romantic dance around the flat for an hour or so.'

'Wow,' I remarked, a neutral placeholder, but Brit was onto me.

She knew me too well, her detective's eyes locking onto mine, scanning for the thing I wasn't saying. But perhaps the puzzle piece she was missing was too far outside the realm of what she could yet grasp. So she let it slide, allowing the conversation to shift as I inquired about her latest gardening creation.

And just like that, she was off, enthusiasm spilling into her hands, sculpting visions in the air, pulling invisible vines from the ether, painting her latest triumph in words thick with colour, with texture, with the breath of achievement.

The weeks tumbled into one another, morning workouts, the rush of London pressing in from every direction, the museum pulling me into its cool, deliberate hush, the book whispering from the edges of my mind, waiting, always waiting. I chipped away at it when I could, fingers brushing over the keys in lunch hour solitude, words coming in fits and starts but coming all the same. Evenings folded over me, the hush of my room thick with the weight of pages unwritten. I told Brit I had a lot going on work-wise, practically becoming a hermit in my room.

Still, I surfaced now and then, a night out here and there, the bright hum of city lights, the slow swell of music in dim-lit spaces, laughter curling like smoke. And always, the anchor of home: Brit, Paul, meals shared over easy conversation, the comfortable give and take of it, the world held at bay for just a little while.

TR and I wrote back and forth, threads of thought stretching across the distance, chapters arriving like waves lapping at the shore. Much of it, we had already unravelled, thread by thread, in our earlier exchanges, familiar yet new, shaped by our long conversations, our in-depths, as I had come to think of them, but it was still satisfying to see everything coming together. Then there were moments, fleeting but powerful, where a surge of Love rushed through me, lifting me past the grind of it all, carrying me forward, weightless, breathless. And then, just like that, the first draft was there, whole, real, staring back at me as if it had always existed.

The forum had quietened, or perhaps I had. A few posts here, a thought there, nothing much, but enough to feel the thread unbroken. And then, a message. Anna.

She knew. The book, near completion. TR had told her.

She wrote:

Hi, Sister,

The circle we have shared in this forum will now close, as the time has come to move quietly to a more protected space. The path ahead calls for stillness, focus, and safety. TR advised me that the book nears its final form, this choice was made not in haste, but in wisdom.

He has spoken with the strength of iron that was forged from the heart, promising to do all he could to help the reservations. He sees what lies dormant beneath, the sacred springs, the quiet will of the children, the forgotten wisdom of the Elders. And he will do all he can to help these places bloom again. Not just as monuments, but as living sanctuaries.

Mitakuye Oyasin.

This is the first step of the great vision of Tȟašúŋke Witkó-Crazy Horse.

Blessings and Love,

Anna

I replied:

Hi, dear Anna,

Alright then, we're going underground. A quiet choice, a wise one, the group doesn't need unwanted attention further on. His word is enough; he will do what he can, where he can, in the ways that matter.

Love,

D

Not long after, I sent TR the manuscript along with an email:

Hi TR,

Anna sent me a message. Closing the forum, a good decision. Best to keep it tucked away now, quiet, protected. She spoke of the reservations, and it still staggers me how, in a nation so advanced, so many are left to live in conditions that shouldn't exist, worlds apart from the wealth that surrounds them. Crazy Horse's vision unfolds at last. Most would dismiss visions as nothing more than mysticism, fanciful imaginings, but like the Lakota, I take them seriously. Some things move beneath the surface, shaping the world whether we see them or not.

Anyway, the draft is done, attached and waiting. It's been a deeply compelling journey, layered, intricate, and relentless from beginning to end. And still, I find myself in awe of your abilities, despite your insistence that everything must remain grounded, rational, and tethered to logic. But even reason has its limits. Some truths simply are. They reveal themselves in their own time.

I have no doubt this book will set the wheels turning, a force in motion that cannot be undone, guiding the course towards something greater, something better, for all of humanity.

Yours sincerely, with Love,

D

I hit send and retreated to Brit's deck with a cup of tea, where the still night air greeted me like an old friend, mingling with the distant hum of traffic and the soft purr of Puddles curled on my lap. The world felt momentarily suspended, wrapped in that quiet hush of midnight, no clatter of thought, just a steady undercurrent of something ancient, something timeless, whispering through the dark.

The draft was done. And with it came a feeling I hadn't quite known before, a quiet certainty, a deep sense of purpose settling into my bones. I drew a slow breath, exhaling accomplishment, letting it sink in.

At the bottom of my teacup, sleep called. I rose, moving towards bed, but as I passed my desk, the dim glow of the screen caught my eye. TR's reply was already there, waiting.

It read:

Hi D,

Yes, best to keep the forum out of sight. Like myself, I will remain anonymous for obvious reasons, and I have associates acting as decoys so I can focus on what truly requires my attention.

Well, the U.S. government broke almost every one of the 368 treaties it signed with Native tribes; legality, it turns out, only moved in one direction. Skirmishes and rivalries existed amongst Indigenous nations, as they do in all human histories, something often cited as a justification for European conquest. But the deeper truth is hard to ignore: it wasn't the tribal way of life that violated those agreements. It was the hand that wrote them. The law, drafted in European terms, was undermined almost entirely by those who claimed to uphold it. What was offered as peace was later enforced as conquest. And what followed was not the spread of civilisation, but the systematic dismantling of one.

If we are serious about solving the world's deeper issues, we must acknowledge what Crazy Horse understood long ago: "*One does not sell the earth upon which the people walk.*" But to get there, to undo centuries of damage, to find real balance, will require a stronger collective intelligence over many generations. That kind of thinking demands a higher intelligence, one capable of recognising the absurdity of drawing rigid lines across a planet we all share. Borders, drawn by blood and madness, could be reshaped through understanding, softened, until the divisions fade into something whole, into a unified solution.

If we fail to move in that direction, if we keep choosing tribalism and greed over wisdom, then we fail the most fundamental test of human intelligence.

As for visions, under normal circumstances, I would be sceptical. But in my youth, I had three, clear, vivid, that were strikingly clear and accurate. Since then, I have kept an open mind. Some visions, the ones that come across as metaphorical, might be the mind grappling with the struggle to decode information from dimensions beyond our grasp. Others reveal obvious trajectories, patterns that deeper intelligence brings to the surface. Perhaps that is why certain songs feel as though they were written. Before the singer even knew the words.

But what people need now isn't mysticism; it's coherence, clear, solid facts, the kind of clarity that cuts through the fog. History repeats itself, and as I write this, the clock ticks 89 seconds closer to midnight. The real crisis isn't just political or environmental; it's the sickness of detachment, tightening its grip on humanity, the underlying force behind centuries of chaos. This book, like my abilities, is part of something much bigger. Destiny is not fate. It is the sum of choices made with open eyes. Truth is the only way.

D, from the depths of my heart, thank you. You've surpassed all expectations in helping with this book, even with so little time to give. Your support has been a gift beyond measure.

Truth and love,

TR

The End

This book ends. But the story doesn't.

If anything, it's only just beginning.

This theory, this alignment, isn't the conclusion of a manuscript. It's the start of a redirection: a shift in how we think, build, and reach. Scientific, humanitarian, environmental, and ethical. None of these domains exists in isolation. They unify, or they collapse.

What we need now is a return to the path we should have taken. A way back to the pulse, we've forgotten how to feel. This is not mysticism. It's not new age. It's old school reborn.

So maybe it's time we stop chasing shadows down theoretical cul-de-sacs and start listening, really listening, to the structure beneath our feet. There are better ways to explore the universe. And to embrace existence. More grounded. More coherent. What if even a fraction of our collective effort were redirected toward something that aligns?

A nonprofit is beginning to take shape under the name Resolutions. Its first aim: to reawaken Berlin's sleeping giant, the International Congress Centre. Not as a monument, but as a mechanism. Not another institution, but a place where minds that still know how to ask real questions can gather. To seek coherence. To find new forms of resolution.

A place for synthesis.

Not for one nation. For one planet.

If the ICC slips beyond reach, the goal remains. There will be a place. Somewhere. Built on the same principles. The project is still forming. But then, so is the future.

If something stirs as you read this, an instinct to build, to help, to bear witness, step aboard.
resolutions.world - This site is a work in progress and will evolve alongside the project. If you'd like to support the early scaffolding of this vision, no pressure, no expectations, you can do so here:

Donate: donorbox.org/resolutions-world

Or directly via PayPal: donate@resolutions.world

www.ingramcontent.com/pod-product-compliance
Lightning Source LLC
LaVergne TN
LVHW091524200926
848483LV00002B/301
* 9 7 8 1 7 6 4 1 1 2 1 1 6 *